SpringerBriefs in Computer Science

More information about this series at http://www.springer.com/series/10028

Dirk Draheim

Generalized Jeffrey Conditionalization

A Frequentist Semantics of Partial Conditionalization

 Springer

Dirk Draheim
Large-Scale Systems Group
Tallinn University of Technology
Tallinn
Estonia

ISSN 2191-5768   ISSN 2191-5776   (electronic)
SpringerBriefs in Computer Science
ISBN 978-3-319-69867-0   ISBN 978-3-319-69868-7   (eBook)
https://doi.org/10.1007/978-3-319-69868-7

Library of Congress Control Number: 2017956747

Printed on acid-free paper

This Springer imprint is published by Springer Nature
The registered company is Springer International Publishing AG
The registered company address is: Gewerbestrasse 11, 6330 Cham, Switzerland

Foreword

In my view, mathematics of the 21st century can be characterized by the attempt at *automating* mathematical reasoning. By Gödel, we know that this will never be completely possible: The higher we go in the sophistication of automation the more remote is the horizon of where we would like to go next. However, it is possible and exciting to conquer higher and higher levels of automation.

The 20th century was the century of *formalizing* mathematical reasoning, which is the first step towards *automating* mathematical reasoning. As a side-product of mathematical formalization, the notion of "universal computer" – which in essence is a mathematical and not an engineering concept – was invented. The enormous impact of this notion on all aspects of science, technology, economy, and society as a whole, by now, is understood by everybody. The impact of *automating* mathematical reasoning (mathematical invention and mathematical verification) will generate bigger and bigger waves of understanding the world and of societal transformation. The waves will include such theoretical areas like, for example, the build-up of web-accessible global and comprehensive mathematical knowledge bases and such practical effects like, for example, deriving hidden knowledge from social media messages.

The level of formalization is not equally high in all areas of mathematics. In this book, Dirk Draheim lays the ground for the formalization of an important part of mathematics that also has high relevance to modern data science: probabilistic reasoning. He clarifies the frequentist semantics of the fundamental notion of partial conditional probability and reveals the subtle differences and the relation between this frequentist and the established Bayesian view. This is the first time that the many results that are due to earlier publications in this area are brought into a coherent form. The concepts can be made operational in today's standard programming paradigms. Thus, the foundational results are immediately available also for practical probabilistic modeling, which is of course of high relevance in current data science and artificial intelligence.

I wish this book wide distribution both in the research community and in the business world.

Research Institute for Symbolic Computation
Johannes Kepler University
Linz/Hagenberg, October 2017   *Bruno Buchberger*

Preface

Statistics is the language of science; however, the semantics of probabilistic reasoning is still a matter of discourse. In this book, I provide a frequentist semantics for conditionalization on partially known events. The resulting frequentist partial (F.P.) conditionalization generalizes Jeffrey conditionalization from partitions to arbitrary collections of events. Furthermore, the postulate of Jeffrey's probability kinematics, which is rooted in Ramsey's subjectivism, turns out to be a consequence in our frequentist semantics.

I think the book appeals to researchers that are involved in any kind of knowledge processing systems. F.P. conditionalization is a straightforward, fundamental concept that fits our intuition. Furthermore, it creates a clear link from the Kolmogorov system of probability to one of the important Bayesian frameworks. This way, I think it is interesting for anybody who investigates semantics of reasoning systems. The list of these mutually overlapping theories, methods and tools includes, without preference, multivariate data analysis, Bayesian frameworks, fuzzy logic, many-valued logics, conditional logic, Nilsson probabilistic logic, probabilistic model checking and also current efforts in unifying probability theory and logics such as the current rational programming.

Tallinn, August 2017   *Dirk Draheim*

Contents

Basic Formulary and Notation 87

Technical Lemmas and Proofs 89

References 95

Index 103

Chapter 1
Introduction

This book provides a frequentist semantics for conditionalization on partially known events which is given as a straightforward generalization of classical conditional probability via so-called probability testbeds. For this purpose, we compare it with an operational semantics of classical conditional probability that is made precise in terms of sequences of so-called conditional events and accompanied by a corresponding instance of the strong law of large numbers. We analyze the resulting partial conditionalization, that we call frequentist partial (F.P.) conditionalization, from different angles, i.e., with respect to partitions, segmentation, independence, and chaining. It turns out that F.P. conditionalization generalizes Jeffrey conditionalization from partitions to arbitrary collections of events, this way opening it for reassessment and a range of potential applications. A counterpart of Jeffrey's rule for the case of independence holds in our frequentist semantics. We compare this result to Jeffrey's commutative chaining of independent updates and the corresponding possible worlds' belief function. Furthermore, the postulate of Jeffrey's probability kinematics, which is rooted in Ramsey's subjectivism and which can be shown analytically equivalent to Donkin's principle, turns out to be a consequence in our frequentist semantics. This way, the book bridges between the Kolmogorov system of probability and one of the important Bayesian frameworks. Then, we will see that an alternative preservation result, i.e., for conditional probabilities under all updated events, holds in our frequentist semantics and exploit it to discuss a possible redesign of the axiomatic basis of probability kinematics. Furthermore, the book looks at desirabilities, which are again a central concept in Ramsey's subjectivism and Jeffrey's logic of decision, and proposes a more fine-grained analysis of desirabilities a posteriori.

The book takes probabilistic reasoning as the subject of investigation. In the past decades, we have seen immense interest in probabilistic reasoning techniques, just think of the artificial intelligence and the data mining community. The book aims to build a path of mitigation between the Bayesian world view and the frequentist world view by giving a frequentist semantics to partial conditionalization. Our approach is reductionist. We take a single, important Bayesian notion as our starting point, i.e., Jeffrey conditionalization by Richard C. Jeffrey [79, 81–87, 89, 92].

© The Author(s) 2017
D. Draheim, *Generalized Jeffrey Conditionalization*, SpringerBriefs in Computer Science, https://doi.org/10.1007/978-3-319-69868-7_1

1.1 From Conditional Probability to Partial Conditionalization

We give a frequentist semantics of conditionalization on arbitrary many partially known events. It turns out that in the special case of non-overlapping events our semantics meets Jeffrey conditionalization. It could be said that we achieve two things, i.e., a generalization of Jeffrey conditionalization plus a pure frequentist interpretation of partial conditionalization. To get the point, first consider the classical notion of conditional probability. Given events A and B, we know that the conditional probability $P(A|B)$ is defined as

$$\boxed{P(A|B) = P(AB)/P(B)} \tag{1.1}$$

The value $P(A|B)$ is called the conditional probability of A under condition B [95]. Now, what is $P(A|B)$ intended to mean? One way to understand it is as follows. The event B has actually occurred, i.e., we have actually observed the event B. Now, $P(A|B)$ expresses the probability that event A has also occurred.

Now, we could say that $P(A|B)$ expresses the idea that the probability of B changes from an old probability $P(B)$, which is, in general different to 100%, into a new probability of 100%. Here, the old probabilities $P(AB)$, $P(A\overline{B})$, $P(\overline{A}B)$, $P(\overline{AB})$ etc. can be called *a priori* probabilities, whereas the new 100%-probability of B and the new $P(A|B)$-probability of A can be called *a posteriori* probabilities.

Now, why allowing the *a priori* probability of the condition B of a conditional probability $P(A|B)$ to be changed into a 100% probability only? Why not allowing it to change into an arbitrary new probability b? Allowing this is exactly what a non-classical conditional probability might be about and what we want to call a partial conditionalization in the sequel. Given a list of events $B_1, \ldots, B_m$ and a list of *a posteriori* probabilities $b_1, \ldots, b_m$, we introduce the notion of probability of event A conditional on the *a posteriori* probability specifications $B_1 \equiv b_1, \ldots, B_m \equiv b_m$ and introduce the following notation for it:

$$P(A \,|\, B_1 \equiv b_1, \ldots, B_m \equiv b_m) \tag{1.2}$$

It is Richard C. Jeffrey who investigates conditional probabilities of the form in Eqn. (1.2) and gives concrete probability values to them, albeit he uses a different notation for them that we will discuss later. He considers those situations, in which the events $B_1, \ldots, B_m$ form a partition of the outcome space. In these cases, Jeffrey gives the following value to partial conditionalizations:

$$\boxed{P(A \,|\, B_1 \equiv b_1, \ldots, B_m \equiv b_m)_J = \sum_{\substack{i=1 \\ P(B_i) \neq 0}}^{m} b_i \cdot P(A \,|\, B_i)} \tag{1.3}$$

The semantics for partial conditionalization expressed by Eqn. (1.3) is known as Jeffrey conditionalization and often also called Jeffrey's rule. We have marked the conditionalization in Eqn. (1.3) with a J as index to distinguish it from the our general notion of partial conditionalization in Eqn. (1.2). Actually, we want to exploit the notation $P(A \mid B_1 \equiv b_1, \ldots, B_m B_m \equiv b_m)$ for any kind of partial conditionalization, depending on the context, and so we use it for our own semantics of partial conditionalization, called F.P. conditionalization, in due course.

What can be the intended meaning of a partial update specification $B \equiv b$? Jeffrey explains this with the notion of *degree of belief* and justifies the Eqn. (1.3) by so-called probability kinematics. We take a different approach. We will give a frequentist semantics to $P(A \mid B_1 \equiv b_1, \ldots, B_m \equiv b_m)$. In a sense, we just generalize the notion of classical conditional probability $P(A|B)$. Let us have a look at $P(A|B)$ again. We have said that we might understand it as follows. B has actually occurred and $P(A|B)$ stands for the probability that A has also occurred. But what can this practically mean? We can think of it as follows. We conduct the original experiment over and over again, i.e., a sufficiently large number of times, and then throw away all the completed experiments in which event B did not occur. Throwing away just means that we ignore them as if they have not occurred. So, we consider only those completed experiments, in which B occurred. Among those, we can expect that event A occurs $P(A|B)$ times. Why? Intuitively, because $P(B)$ is the expected relative number of occurrences of B and, similarly, $P(AB)$ is the expected relative number of occurrences of AB, so that we can expect $P(AB)/P(B)$ as the relative number of occurrences of A in those completed experiments that yielded B. This intuition adheres completely to Andrey Kolmogorov's original explication of probability theory in [95], which characterizes a probability value as the frequency of an event in a repeated experiment. Formally, this understanding of the conditional probability $P(A|B)$ is backed by the laws of large numbers. And it is exactly this understanding of the conditional probability that guides us in giving a frequentist semantics to the partial conditionalization $P(A \mid B_1 \equiv b_1, \ldots, B_m \equiv b_m)$.

Now, how do we give semantics to $P(A \mid B_1 \equiv b_1, \ldots, B_m \equiv b_m)$? We consider repeated experiments of such lengths, in which statements of the form $B_i \equiv b_i$ make sense frequentistically, i.e., the probability b_i can be interpreted as the frequency of B_i and can potentially be observed. Then we reduce the notion of partial conditionalization to the notion of classical conditional probability. We consider the expected value of the frequency of A, i.e., the average occurrence of A, conditional on the event that the frequencies of events B_i adhere to the new probabilities b_i. Now, we can speak of the b_is as frequencies. We can speak of pairs $B_i \equiv b_i$ also as frequency specifications or probability specifications. We call the resulting notion frequentist partial conditionalization (F.P.) conditionalization. Now, how do we write down, what we just have said, in standard notation? Given an adequate number n of experiment repetitions, we will define the F.P. conditionalization bounded by n and denoted by $P^n(A \mid B_1 \equiv b_1, \ldots, B_m \equiv b_m)$ as follows:

$$P^n(A \mid B_1 \equiv b_1, \ldots, B_m \equiv b_m) = \mathsf{E}(\overline{A^n} \mid \overline{B_1^n} = b_1, \ldots, \overline{B_m^n} = b_m) \qquad (1.4)$$

In general, the F.P. conditionalization $P(A \mid B_1 \equiv b_1, \ldots, B_m \equiv b_m)$ will be defined via the limit of expected values in all repeated experiments of adequate numbers of repetitions. Then bounded F.P. conditionalizations $P^n(A \mid B_1 \equiv b_1, \ldots, B_m \equiv b_m)$ of the form Eqn. (1.4) are approximations to an F.P. conditionalization. In some important cases, the F.P. conditionalizations $P^n(A \mid B_1 \equiv b_1, \ldots, B_m \equiv b_m)$ are equal for all adequate numbers of repetitions n. First, this is so in case of Jeffrey conditionalization, in which the events B_1 through B_m form a partition of the outcome space. Second, this is so in case the events B_1 through B_m are mutually independent.

Remarks on Notation and Terminology

Jeffrey also gives a value to the probability of an event A conditional on a single probability specification $B \equiv b$ as follows:

$$P(A \mid B \equiv b)_J = b \cdot P(A \mid B) + (1 - b) \cdot P(A \mid \overline{B}) \qquad (1.5)$$

Equation (1.5) can be considered as an instance of Eqn. (1.3) due to the fact that $P(A \mid B \equiv b)_J$ can be rewritten as $P(A \mid B \equiv b, \overline{B} \equiv (1 - b))_J$. Equation (1.5) itself is also often called Jeffrey conditionalization or Jeffrey's rule.

The notation for partial conditionalization $P(A \mid B_1 \equiv b_1, \ldots, B_m \equiv b_m)$ is an arbitrary choice, in particular, the notation for probability specifications $B \equiv b$. A probability specification $B \equiv b$ expresses the fact that the probability of an event B changes from an *a priori* probability $P(B)$ to a new, *a posteriori* probability b. So, maybe some notation such as $P(B) \leadsto b$ or $P(B) := b$ would be more appropriate. We have chosen the short notation $B \equiv b$ for the sake of readability.

Note that our notation for Jeffrey conditionalization $P(A \mid B_1 \equiv b_1, \ldots, B_m \equiv b_m)_J$ in Eqn. (1.3) is oriented towards our own notation for partial conditionalizations. The original notation of Jeffrey in [87] is different. Jeffrey denotes all *a priori* probabilities as $prob(A)$ and all *a posteriori* probabilities as $PROB(A)$ so that Eqn. (1.3) looks like

$$PROB(A) = \sum_{\substack{i = 1 \\ P(B_i) \neq 0}}^{m} PROB(B_i) \cdot prob(A \mid B_i) \qquad (1.6)$$

In particular, Jeffrey does not make explicit the conditions of $PROB(A)$ as we do with the $B_i \equiv b_i$ in $P(A \mid B_1 \equiv b_1, \ldots, B_m \equiv b_m)$. Rather, Eqn. (1.6) relies on the context making clear under which conditions $PROB(A)$ is a conditionalization. Actually, the question of explicit or implicit notation has no semantic implications. With respect to Jeffrey conditionalization and other kinds of possible world conditionalizations, like the one considered by William F. Donkin, this question is a merely notational issue. Actually, in early literature Jeffrey uses yet another option halfway between the implicit and the explicit notations, with explicit update values $b_1, \ldots, b_m$ that are maintained in the context. For F.P. conditionalization, the explicit notation is essential. We define the conditionalization, we do not simply postulate it. Therefore, we need to proof first that the *a posteriori* probability $P(B_i \mid B_1 \equiv b_1, \ldots, B_m \equiv b_m)$ of an event B_i equals the value b_i that we assign to it. In possible world frameworks, this is not a question; it is just taken per establishment of the conditionalization. Anyhow, $P(B_i \mid B_1 \equiv b_1, \ldots, B_m \equiv b_m)$ is the first thing that we will proof for F.P.

semantics and then could use an implicit notation, at least in large parts, also for F.P. conditionalization. Again, the question of explicit vs. implicit notation does not matter conceptually, in particular, when it comes to comparisons between F.P. semantics and Jeffrey conditionalization or other possible world semantics.

When we deal with F.P. conditionalization, we only use the explicit notation. The explicit notation shows a neat correspondence between the frequency specifications $B \equiv b$ and $\overline{B^n} = b$. Also, the explicit notation is easier to maintain in proofs and argumentations. When we deal with Jeffrey conditionalization or other conditionalizations from the Bayesian realm as, e.g., found in Donkin's principle later, we feel free to use either the implicit or the explicit notation. Usually, we want to switch to the explicit notation when it comes to comparisons between different conditionalization frameworks.

1.2 Background, Motivation, Perspectives and Outlook

We make conditionalization over partial update specifications the subject of our interest. We do not want to take a position in favor or against one of the many important logical apparatuses, reasoning frameworks. We even try to avoid statements in favor or against one of the big schools, the Bayesian or the frequentist. Rather, we want to provide a bridge between the Bayesian worldview and the frequentist world. F.P. conditionalization has the following key characteristics. F.P. conditionalization semantics if consistent with the original explication of probability theory as characterized by Andrey Kolmogorov. It makes no use of further concepts from psychology, epistemology, decision theory etc. F.P. conditionalization turns out to be consistent with Jeffrey conditionalization, i.e., we can prove that it yields the same values as Jeffrey conditionalization – in the special case that is treated by Jeffrey conditionalization, i.e., partitions. F.P. conditionalization is a generalization of Jeffrey conditionalization. Jeffrey conditionalization is defined for partitions only. F.P. conditionalization is different. It is defined, declaratively, for arbitrary lists of events. Jeffrey conditionalization relies on the postulate that an *a priori* conditional probability with respect to one of the updated events remains unchanged after conditionalization, i.e., after update of the condition event. The justification of Jeffrey's rule needs this so-called kinematics postulate. This is different in the F.P. framework. Here, Jeffrey's rule is a consequence of the definition of F.P. semantics. F.P. conditionalization creates a link from the Kolmogorov system of probability to one of the important Bayesian frameworks, i.e., Jeffrey's logic of decision.

The approach of this book is formal. We take a notion from the Bayesian realm, i.e., conditionalization with respect to partial updates, and will equip it with a formal, rigor semantics from the frequentist realm. We hope that the results are useful for the reader for his or her own reception, in his or her own, individual assessment of reasoning techniques and approaches.

Frequentism can be clearly identified with what Julian Jaynes calls the Kolmogorov system of probability [78]. Here, we are stable and safe, I mean from a technical viewpoint. The frequentist approach is the perspective of Bernoulli and his Golden Theorem. It has an authoritative explication, the observation of a repeated experiment in a long series of repetitions [94–96]. Once formalized, e.g., in today's standard formalization in measure theory, this explication is reinforced or let's say confirmed by the laws of large numbers and the central limits theorems that follow from this formalization. This way, the frequentist viewpoint emerged into a particularly consistent and strong argumentative framework. So, when it comes to frequentism, we have a very clear understanding of what this is about. Here, the question of objectivism is not relevant for the developments in the book, and we rather would like to refer to the discussions conducted by Jerzy Neyman [116, 118] with respect to this. We use frequentism rather as synonym for the Kolmogorov theory together with its standard explication. Once we have agreed that this is not an oversimplification, it is fair to say, again, that we have a very clear understanding of what frequentism is about.

When it comes to Bayesianism, things are different. There is no such single, closed apparatus as with frequentism. Instead, as you know, there is a great variety of important approaches and methodologies, with different flavors in objectives and explications [64, 153, 155]. Think of Bruno de Finetti [58, 59] with his Dutch book argument and Frank P. Ramsey [127, 129] with his representation theorem [128]. Think of Julian Jaynes [76], who starts from improving statistical reasoning with his application of maximal entropy [77], and from there transcends into an agent-oriented explanation of probability theory [78]. Also, think of Judea Pearl [123], who eventually transcends probabilistic reasoning by systematically incorporating causality into his considerations [124, 125]. Bayesian approaches have in common that they rely, at least in crucial parts, on notions other than frequencies to explain probabilities, among the most typical are degrees of belief, degrees of preference, degrees of plausibility, degrees of validity, degrees of confirmation or, on the opposite side, degrees of uncertainty. Also, they have in common, that they, more or less, stress and exploit notions of probability update, typically traced back to Bayes' rule and then usually called Bayesian update.

Let us have a look at Rudolf Carnap's Probability–1 and Probability–2 [23, 24] that he uses to distinguish between two fundamentally different concepts of probability. It is Probability–2 that refers the frequentist interpretation of probability theory. Now, Carnap's Probability–1 is clearly about degree of confirmation. But there are more characteristics that distinguish Probability–1 from Probability–2 and it is interesting to have a look at them. For Carnap, Probability–2 is clearly an empirical concept; it is about observation of events. And sentences in Probability–2 are sentences about factual states of affairs. And now, with Probability–1 sentences are not about observations, there are considered purely analytical entities. He uses terms such as semantical and logical to characterize this notion of analytical probability description.

All the single important Bayesian frameworks are each highly sophisticated and elaborated. However, as we said, Bayesianism does not stand for a single closed framework, but rather for a variety of frameworks that share, more or less, some common characteristics. In this book, we bridge between frequentism and one of the important Bayesian frameworks, i.e., the logic of decision of Richard C. Jeffrey. First, we give our frequentist semantics to a general notion of partial conditionalization. Then it turns out that our partial conditionalization semantically meets Jeffrey conditionalization, which is the core concept of Jeffrey's probability kinematics, which is again a or even *the* crucial building block of Jeffrey's logic of decision. It turns out that the found concept is, in a sense, is a true generalization of Jeffrey conditionalization, as Jeffrey conditionalization is defined only for partitions, i.e., collections of condition events that form a partition. As opposed to that, the found partial conditionalization works for arbitrary, also overlapping, condition events.

In probability kinematics , a certain property of probabilities and expectations is assumed as given. It is postulated. It is the invariance of a conditional probability of a target event over one of the condition events *a posteriori*, i.e., after conditionalization. The postulate is crucial for probability kinematics for two reasons. First, it is needed to justify the definition of Jeffrey conditionalization. Actually, the definition of Jeffrey conditionalization follows from the postulate via the law of total probability. Second, and equally important, it brings Jeffrey's logic of decision together, conceptually, with the important Bayesian framework of Frank P. Ramsey. Now, our frequentist conditionalization sheds some new light on the postulate of probability kinematics. In our frequentist semantics the postulate follows from the definition of our conditionalization, whereas, in the logic of decision, it is assumed as a basic concept, as the axiomatic basis so to speak. Furthermore, it turns out, that the special case of partitions is essential to probability kinematics, as Jeffrey conditionalization can only be derived from its postulate and therefore justified if the events are guaranteed to form a partition.

When I first dealt with F.P. conditionalization, I was aware of Jeffrey conditionalization; however, my efforts were not particularly intended towards Jeffrey conditionalization. The idea was to generalize classical conditional probabilities, so that partially known events make sense frequentistically. The idea was then to look at sequences of base experiments in which partial probability specifications can be observed and this way make sense of partially known events. Now, the question was what we can expect as the relative occurrence of a target event in such probability testbeds. When I first dealt this notion, I was immediately thinking about it in terms of its combinatorial solution, see Sect. 4.1.2, i.e., a true frequentist approach so to speak. The combinatorial solution of the involved multivariate Bernoulli distributions is a straightforward generalization of the combinatorial solution for basic Bernoulli distributions. Declaratively, the searched concept is the conditional expected value over partial update specifications, and this is how our notion of partial conditionalization is defined.

In this book we are interested in probabilistic reasoning in its own right. In [44] we are interested in reasoning about systems that contain programmable components and act in a probabilistic environment. We take a reductionist approach in [44] and investigate the probabilistic typed lambda-calculus, which is a minimal functional programming language plus a probabilistic programming choice construct. The work in [44] is motivated by previous work on the engineering and operation of business process technology [5, 40, 42, 48–50]. Given the relevance of probabilistic reasoning to business processes, compare with [25, 29, 30], this closes the loop to the current book. In [43] we deal with reflective constraint writing, which is the counterpart of full reflective programming with respect to constraint writing. Reflective constraint writing has its application in multi-level systems engineering and maitenance [4, 11, 41, 45–47, 51, 52, 108], however, its original motivation was in semantic clarification of multi-level modelling [53, 54, 70]. With the investigated decoupling of the intention vs. the pragmatics of multi-level modeling languages the work touches the considerations in the current book.

1.3 Chapter Overview

In Chap. 2 we define F.P. conditionalization. We start the chapter with a recap on how to model repeated experiments. Then we define F.P. conditionalization as the expected value of an event in a testbed that adheres to frequency constraints. It also presents, as a Lemma, an alternative definition of F.P. conditionalization as a conditional probability with respect to adequate testbeds. The second definition is easier to handle in proofs and argumentations and is, therefore, the *de facto* definition of F.P. conditionalization on many occasions. The chapter proceeds with looking into a particularly intuitive kind of F.P. conditionalizations, i.e., such that are projective in the target event. Next, the chapter looks into an important technique for transforming F.P. testbed specifications, i.e., shortening a testbed while adjusting the frequencies of the involved frequency specifications. Finally, we look into an operational semantics of classical conditional probabilities resulting into the notion of so-called conditional event and the corresponding instance of the strong law of large numbers.

Chapter 3 deals with classical Jeffrey conditionalization. Jeffrey conditionalization deals with the special case of condition events that form a partition. We show that F.P. conditionalization meets Jeffrey conditionalization. We start with the basic case of a single condition event and proceed with Jeffrey conditionalization in general. With Chap. 3 Jeffrey conditionalization is technically embedded into F.P. conditionalization.

Chapter 4 investigates properties of full, general F.P. conditionalizations. We look into how F.P. conditionalization can be computed. We do so by giving both a recursive definition as well as a combinatorial characterization. Next, we present a segmentation Lemma which draws a further, convenient analog to the special case of Jeffrey conditionalization. Next, we will see that independence of condition events is preserved by F.P. conditionalization. As a consequence, it is easy to determine the value of each F.P. conditionalization in case of independent conditions. The independence result is important as it again reinforces our intuition about F.P. conditionalization. Next, we will see in how far classical conditional probabilities *a priori* are preserved under F.P. update. Furthermore, we investigate the behavior of expected values under after F.P. conditionalization.

In Chap. 5 we conduct a discussion of Jeffrey's probability kinematics and F.P. semantics. We discuss Jeffrey's postulate and compare Jeffrey's commutative chaining of independent events with F.P. segmentation of independent events. Against this background, we discuss the redesign of the axiomatic basis of probability kinematics. Furthermore, we will discuss the subjectivist concept of desirabilities and propose a fine-grained investigation of desirabilities after partial update. Furthermore, we will investigate the correspondence between Donkin's principle and Jeffrey's postulate. It turns out that Donkin's principle and Jeffrey's postulate are equivalent.

Appendix A provides bibliographic notes on related work with respect to Richard C. Jeffrey's writings, contingency tables for given observations and marginals, rationales of probability kinematics, closeness of probability measures, non-commutativity of Jeffrey conditionalization, Dempster-Shafer theory, the maximal entropy principle, Jeffrey desirabilities and a series of further relevant topics. Appendix B serves as a repository for some mathematical definitions, whereas Appendix C compiles some technical Lemmas that are important in proofs and argumentations of the book.

Chapter 2
F.P. Conditionalization

In this chapter, we introduce the notion of frequentist partial conditionalization, F.P. conditionalization for short. An F.P. conditionalization is a probability conditional over a list of event-probability specifications, and we introduce the notation $P(A \mid B_1 \equiv b_1, \ldots, B_m \equiv b_m)$ for it. A specification pair $B \equiv b$ stands for the assumption that the probability of event B has somehow changed from a previously given probability $P(B)$ into a new probability b. Now, we give a frequentist explanation of the probability specifications. We think of conditionalization as taking place in chains of repeated experiments, so called probability testbeds. The semantics is introduced independent of psychological metaphors such as degrees of belief. It also comes without any kind of possible worlds semantics. On the contrary, with the F.P. semantics, there is an underlying probability space that remains over all the time, albeit we consider new, additional probability functions after each update.

With F.P. conditionalization, we generalize Jeffrey conditionalization. Jeffrey conditionalization treats the special case of completely decomposing conditions, i.e., the case in which the condition events B_1 through B_m form a partition of the outcome space. With our semantics, this requirement can be dropped so that we can deal with arbitrary lists of overlapping events. To give a first impression, let us have a look at a series of cases in which F.P. conditionalization falls together with classical conditional probability. First, if the conditions are specified as 100%, F.P. conditionalization just amounts to an ordinary conditional probability, i.e., we have the following:

$$P(A \mid B_1 \equiv 100\%, \ldots, B_m \equiv 100\%) = P(A \mid B_1, \ldots, B_m) \tag{2.1}$$

An F.P. conditionalization is distinguished from the classical conditional probability by the fact that the conditions are given as pairs of the form $B \equiv b$. The fact expressed by Eqn. (2.1) meets our intuition. If we specify a single condition B as having a 100% probability, we might want to express that B has actually occurred. Now, if we have observed that some events B_1 through B_m have actually occurred, we would like to adjust our opinion about the probability of a further, not yet observed event A, to the probability of A conditional under events B_1 through B_m.

© The Author(s) 2017
D. Draheim, *Generalized Jeffrey Conditionalization*, SpringerBriefs in Computer Science, https://doi.org/10.1007/978-3-319-69868-7_2

Similarly, in case we set all conditions to a 0% probability, our semantics of F.P. conditionalization again yields a classical conditional probability, albeit under the complements of the conditions this time. Again this meets our intuition completely:

$$P(A \mid B_1 \equiv 0\%, \ldots, B_m \equiv 0\%) = P(A \mid \overline{B_1}, \ldots, \overline{B_m}) \qquad (2.2)$$

Next, if we select the original probability of a single condition event as its new probability, we would expect that F.P. conditionalization has no effect at all. And actually, this is what our semantics will yield, i.e.:

$$P(A \mid B \equiv P(B)) = P(A) \qquad (2.3)$$

We proceed as follows. With Sect. 2.1 we start with a technical, preparatory section that compiles relevant definitions, concepts and notation. However, Sect. 2.1 should not be skipped, because some of the discussions in Sect. 2.1 are crucial for the rest of the book. In Sect. 2.2 we will formally define the notion of F.P. conditionalization and further discuss the intuition behind it. With the so-called projective F.P. conditionalization in Sect. 2.3 we will look at a particularly intuitive F.P. conditionalization and show how it is met by our frequentist semantics. In Sect. 2.4 we will discuss some central properties of frequentist testbed specifications. We show how testbed specifications can be shortened by cutting and adjusting frequencies. This technique is an important tool in many argumentations and proofs concerning F.P. conditionalization. In service of deeper analysis of F.P. semantics, Sect. 2.5 discusses an operational semantics of classical conditional probability that is given in terms of a sequence of so-called conditional events and accompanied by a corresponding instance of the strong law of large numbers.

2.1 On Modeling Repeated Experiments

2.1.1 Independent, Identically Distributed Random Variables

As usual, we model repeated experiments as sequences of mutually independent, identically distributed random variables. In this section, we recap the necessary notions. The notion of independence is crucial for reasoning in probabilistic environments. We consider an event A as independent of an event B, if its overall probability does not change, whenever we narrow our observation to its probability conditional on event B, i.e., if we have that

$$P(A) = P(A \mid B) \qquad (2.4)$$

Another, equivalent characterization of independence is the following:

$$P(A \mid B) = P(A \mid \overline{B}) \qquad (2.5)$$

Equation (2.5) is also particularly intuitive. Where Equation (2.4) expresses that A is independent of whether we narrow our focus on the occurrence of B or not, Eqn. (2.5) expresses that A is independent of whether B has occurred or not, once we have narrowed our focus on the occurrence of B. Note that the notion of independence is a completely formal concept. Though its name hints to causality, i.e., the mutual influence of experimental factors, dependencies and independencies of events do not at all necessarily entail some causalities resp. non-causalities. The question, in how far and under which circumstances independence and dependency can be exploited in reasoning about casualties is a subtle issue. Nevertheless, we will use mutually independent events, or random variables to be more precise, to model repeated mutually non-influencing experiments. Given the definition of conditional probability; compare with Eqn. (1.1), we have that Eqn. (2.4) is equivalent to the following:

$$P(A) = P(AB)/P(B) \tag{2.6}$$

Now, given Eqn. (2.6) we also have that $P(B) = P(B|A)$ and therefore, that B is independent of A, whenever A is independent of B. Furthermore, we have that Eqns. (2.4) and (2.6) can be rewritten as follows:

$$P(AB) = P(A) \cdot P(B) \tag{2.7}$$

All characterizations in Eqns. (2.4) through (2.7) express the same notion of independence, i.e., a notion of *observational proportional independence*, where Eqn. (2.7) might be considered closest to this intuition. Actually, Eqn. (2.7) is the form that is often used as the definition of independence *per se* and, similarly, it is the form that we will use in upcoming definitions.

Next, we generalize the notion of independent events to random variables. In upcoming definitions for random variables, we always assume a probability space (Ω, Σ, P) given as context.

Definition 2.1 (Independent Random Variables) Given two discrete random variables $X : \Omega \to I_1$ and $Y : \Omega \to I_2$, we say that X and Y are *independent*, if the following holds for all index values $v \in I_1$ and $v' \in I_2$:

$$P(X = v, Y = v') = P(X = v) \cdot P(Y = v') \tag{2.8}$$

Definition 2.2 (Pairwise Independence) Given a finite list of discrete random variables $X_1 : \Omega \to I_1$ through $X_n : \Omega \to I_n$, we say that $X_1, \dots, X_n$ are *pairwise independent*, if X_i and X_j are independent for any two random variables $X_i \neq X_j$ from $X_1, \dots, X_n$.

Definition 2.3 (Mutual Independence) Given a finite list of discrete random variables $X_1 : \Omega \to I_1$ through $X_n : \Omega \to I_n$, we say that $X_1, \dots, X_n$ are *(mutually) independent*, if the following holds for all index values $v_1 \in I_1$ through $v_n \in I_n$:

$$P(X_1 = v_1, \dots, X_n = v_n) = P(X_1 = v_1) \times \cdots \times P(X_n = v_n) \tag{2.9}$$

Mutual independence implies pairwise independence, but not vice versa. Given mutually independent random variables, they are also called just independent for short. Next, we step from finite lists of random variables to a notion of independence for countable infinite lists of random variables.

Definition 2.4 (Countable Independence) Given a sequence of discrete random variables $(X_i : \Omega \to I_i)_{i \in \mathbb{N}}$ we say that they are *(all) independent*, if for each finite set of indices $i_1, \ldots, i_m$ we have that $X_{i_1}, \ldots, X_{i_m}$ are mutually independent.

Next, we recap the notion of identically distributed random variables. Two random variables that operate on the same probability space and have the same value set are said to be identically distributed if they cannot be distinguished only in terms of their probabilities or, to be more precise, in terms of probabilities they entail.

Definition 2.5 (Identically Distributed Random Variables) Given random variables $X : \Omega \to I$ and $Y : \Omega \to I$, we say that X and Y are *identically distributed*, if the following holds for all $v \in I$:

$$P(X = v) = P(Y = v) \tag{2.10}$$

We need random variables that are both identically distributed and independent to model repeated experiments, i.e., one random variable for each repetition that we want to represent. Therefore, we now introduce the notion of independent identically distributed random variables as an explicit concept, denoted by i.i.d. , i.e., both for pairs of random variables as well as for sequences of random variables of arbitrary length.

Definition 2.6 (Independent, Identically Distributed)   Given two random variables $X : \Omega \to I$ and $Y : \Omega \to I$, we say that and X and Y are *independent identically distributed*, abbreviated as i.i.d., if they are both independent and identically distributed.

Given to identically distributed random variables X and Y, we necessarily have that their expected values are equal, i.e., $E(X) = E(Y)$.

Definition 2.7 (Sequence of i.i.d. Random Variables) Random variables $(X_i)_{i \in \mathbb{N}}$ are called *independent identically distributed*, again abbreviated as i.i.d., if they are all independent and, furthermore, all identically distributed.

In general, each family $(A_i)_{i \in \mathbb{N}}$ of i.i.d. random variables can be thought of as modeling a repeated experiment. Later, we are particularly interested in families of i.i.d. characteristic random variables, because they sum up to representations of absolute and relative occurrences of events in repeated experiments.

2.1.2 Sequences of Multivariate Random Variables

Next, we are interested in observing not only one event but a system of possible events concerning the outcomes of a repeated experiment. Therefore, we would like to look simultaneously at a list of sequences of i.i.d. random variables that should jointly model a repeated experiment. Unfortunately, it is not enough to require that all the single involved sequences of random variables are sequences of i.i.d. random variables. Independence must spawn the several observed events crosswise to model the scenario adequately. Here, care must be taken. With respect to a single repetition of the experiment, we must not require independence, however, with respect to different experiments, all involved events must be independent. Such scenarios are adequately described by sequences of i.i.d. multivariate random variables.

Definition 2.8 (Multivariate Random Variable)   Given a list of n so-called *marginal random variables* $X_1 : \Omega \longrightarrow I_1$ to $X_n : \Omega \longrightarrow I_n$, we define the *multivariate random variable* $\langle X_1, \ldots, X_n \rangle : \Omega \longrightarrow I_1 \times \cdots \times I_n$ for all outcomes $\omega \in \Omega$ as follows:

$$\langle X_1, \ldots, X_n \rangle(\omega) = \langle X_1(\omega), \ldots, X_n(\omega) \rangle \tag{2.11}$$

As a merely notational issue, we have the following relationship between a multivariate random variable and its marginal random variables:

$$\mathsf{P}(\langle X_1, \ldots, X_n \rangle = \langle i_1, \ldots, i_n \rangle) = \mathsf{P}(X_1 = i_1, \ldots, X_n = i_n) \tag{2.12}$$

We can think of a sequence of i.i.d. multivariate random variables as a matrix of marginal random variables, where the multivariate random variables play the role of column vectors. Then, the repeated observation of one of the involved events results from this matrix as a row vector of marginals. Now, column-wise, the random variables of this matrix are, in general, not independent. Despite that, the random variables are all independent as expressed more precisely in Lemma C.1. In particular, the row vectors form i.i.d.sequences of random variables as expressed in Corollary 2.9.

Corollary 2.9 (I.I.D. Multivariate Random Variable Marginals)   *Given a sequence of i.i.d. multivariate random variables* $(\langle X_1, \ldots, X_n \rangle_i)_{i \in \mathbb{N}}$ *we have that all marginal sequences* $((X_1)_i)_{i \in \mathbb{N}}$ *through* $((X_n)_i)_{i \in \mathbb{N}}$ *are i.i.d.*

Proof. Immediate Corollary from Lemma C.1.   □

The converse of Corallary 2.9 does not hold, i.e., given sequences $((X_1)_i)_{i \in \mathbb{N}}$ through $((X_n)_i)_{i \in \mathbb{N}}$ that are all i.i.d. we do not have, automatically, that the sequence of multivariate random variables $(\langle X_1, \ldots, X_n \rangle_i)_{i \in \mathbb{N}}$ is also i.i.d.

2.1.3 Sums and Averages of Random Variables

Given two real-valued random variables $X : \Omega \to \mathbb{R}$ and $Y : \Omega \to \mathbb{R}$ their sum $X + Y : \Omega \to \mathbb{R}$ is defined pointwise for all $\omega \in \Omega$ as follows:

$$(X + Y)(\omega) = X(\omega) + Y(\omega) \tag{2.13}$$

The sum of two random variables is again a random variable. We can characterize the sum of two random variables also via the inverse images it entails. For all $r \in \mathbb{R}$ we have the following:

$$((X + Y) = r) \quad = \quad \{\, \omega \,|\, X(\omega) + Y(\omega) = r \,\} \tag{2.14}$$

The perspective on $X + Y$ given by Eqn. (2.14) is important, because it tells us how to calculate probabilities under $X + Y$ via the decomposition into probabilities under X and under Y, component-wise. For all $r \in \mathbb{R}$ we have the following:

$$\mathsf{P}((X + Y) = r) = \sum_{r_x + r_y = r} \mathsf{P}(X = r_x, Y = r_y) \tag{2.15}$$

Given a sequence of real-valued random variables $(X_i)_{i \in \{1,\dots,n\}}$, their sum is denoted as X^n. Obviously, X^n is defined as $((\cdots(((X_1 + X_2) + X_3) + X_4) + \cdots) + X_n)$.

Convention 1 (Sum of Random Variables from First Position) Given an infinite sequence of real-valued random variables $(X_i)_{i \in \mathbb{N}}$ we use X^n to denote the sum of the first n random variables $X_1 + \cdots + X_n$

Convention 2 (Sum of Random Variables from Arbitrary Position) Given an infinite sequence of real-valued random variables $(X_i)_{i \in \mathbb{N}}$ and a starting position j we use X_j^n to denote the sum $X_j + X_{j+1} + \cdots + X_{j+n-1}$. Obviously, we have that $X^n = X_1^n$.

If the random variables $(X_i)_{i \in \mathbb{N}}$ are i.i.d. we have the following for any two collections of n indices $i_1, \dots, i_n$ and $j_1, \dots, j_n$ and all real numbers r:

$$\mathsf{P}(X_{i_1} + \cdots + X_{i_n} = r) = \mathsf{P}(X_{j_1} + \cdots + X_{j_n} = r) \tag{2.16}$$

With Eqn. (2.16) we can rearrange or shift around a sum arbitrarily. For the result to hold, an identical distribution is not sufficient, because the index sets may overlap. It is crucial that the random variables are also all independent. For our proofs, it will be important that the result can be extended to multivariate random variables. Given an i.i.d. sequence $(\langle X_1, \dots, X_m \rangle_i)_{i \in \mathbb{N}}$ of real-valued multivariate random variables, we have for any two collections of n column indices $i_1, \dots, i_n$ and $j_1, \dots, j_n$ and all real numbers $r_1, \dots, r_m$:

$$\mathsf{P}(\bigcap_{1 \leqslant k \leqslant m} (X_k)_{i_1} + \cdots + (X_k)_{i_n} = r_k) = \mathsf{P}(\bigcap_{1 \leqslant k \leqslant m} (X_k)_{j_1} + \cdots + (X_k)_{j_n} = r_k) \tag{2.17}$$

Eqn. (2.17) expresses that we can rearrange or shift around a series of sums, as long as the moves are made *simultaneously* in all involved marginals, i.e., column-wise. This also explains why Convention 1 and 2 matter. In general, we have that $P(\cap_{1 \leqslant k \leqslant m} X_k^n = r_k)$ does not equal $P(\cap_{1 \leqslant k \leqslant m} (X_k)_{i_{k1}} + \cdots + (X_k)_{i_{kn}} = r_k)$ for arbitrary collections of $m \times n$ indices $(i_{xy})_{\{x \leqslant m, y \leqslant n\}}$, even in case $(\langle X_1, \ldots, X_m \rangle_i)_{i \in \mathbb{N}}$ is i.i.d.

Furthermore, we also define the notion X^0 of a sum of length zero. This is needed later in some Lemmas and proofs about sums of arbitrary length X^n for $n \geqslant 1$ in wich, however, some random variable of the form X^{n-1} occurs. Again, $X^0 : \Omega \to \mathbb{R}$ is a real-valued random variable, which is defined as a constant function for all $\omega \in \Omega$ as follows:

$$X^0(\omega) = 0 \qquad (2.18)$$

With Eqn. (2.18) we have that $(X^0 = 0) = \Omega$ and, consequently, $(X^0 = r) = \emptyset$ for all $r \neq 0$.

Given a random variable $X : \Omega \to \mathbb{R}$ and a real number $r \in \mathbb{R}$ the product $r \cdot X$ is defined pointwise for all $\omega \in \Omega$ and $r, i \in \mathbb{R}$ as follows:

$$(r \cdot X)(\omega) = r \cdot X(\omega) \qquad (2.19)$$

$$P(r \cdot X = i) = P(X = i/r) \qquad (2.20)$$

Given a sequence $X_1, \ldots, X_m$ of random variables we denote their average random variable, or just average for short, as $\overline{X_1 + \ldots + X_n}$ or $\overline{X^n}$. We define $\overline{X^n}$ on the basis of X^n as follows:

$$\overline{X^n} = 1/n \cdot X^n \qquad (2.21)$$

Again, for an infinite sequence of real-valued random variables $(X_i)_{i \in \mathbb{N}}$ we use $\overline{X^n}$ to denote the average of the first n random variables $\overline{X_1 + \cdots + X_n}$. Again, averages for i.i.d. random variables are invariant with respect to moving summands arround, i.e., the counterparts of Eqns. (2.16) and (2.17) hold also for averages.

Last, as an important further property, we have that mutual independence of random variables is preserved by sums, multiplication and averages; compare with Lemma C.4.

2.1.4 Sequences of Characteristic Random Variables

A characteristic random variable is a real-valued random variable $A : \Omega \to \mathbb{R}$ that can take only zero or one as values, i.e., $(A = 1) \cup (A = 0) = \Omega$. Characteristic random variables are also called Bernoulli variables. A characteristic random variable characterizes an event. Given an event $A \subseteq \Omega$ we define its characteristic random variable so that $A(\omega) = 1$ for all $\omega \in A$ and $A(\omega) = 0$ for all $\omega \notin A$. Note that we overload the name of the event A with the name of its characteristic random variable. Vice versa, given a characteristic random variable A, we simply use A to denote the event $A = 1$. A characteristic random variable stands for a Bernoulli experiment and determines a Bernoulli distribution. With respect to the Bernoulli experiment, the event $A = 1$ is called a success. We can use an i.i.d. sequence of characteristic random variables $(A_i)_{i \in \mathbb{N}}$ to model the repetition of an experiment. Usually, we will use $A_{(i)}$ instead of A_i to denote the i-th characteristic random variable. When we use multivariate random variables $A_{j(i)}$, this makes it more convenient to distinguish the j-th component from the i-th repetition compared with the alternative notations $(A_j)_i$ or A_{ji}. We say that $A_{(i)}$ stands for the fact that A occured upon the i-th repetition of the experiment. As a further notational convention, we use A to denote the first random variable $A_{(1)}$. Once more, we use $A_{(i)}$ to denote the event $A_{(i)} = 1$, see above. Therefore, we also use A to denote the event $A_{(1)} = 1$. We have that $P(A^n = k)$ is the probability that for an outcome ω there exists a subset $\{i_1, \ldots, i_k\} \subseteq \{1, \ldots, n\}$ of the first n indices of length k such that $A_{(i)}(\omega) = 1$ for all $i \in \{i_1, \ldots, i_n\}$, and $A_{(i')}(\omega) = 0$ for all $i' \in \{1, \ldots, n\} \backslash \{i_1, \ldots, i_k\}$, i.e.,

$$P(A^n = k) = \sum_{\substack{I = \{i_1, \ldots, i_k\} \\ I' = \{i'_1, \ldots, i'_{n-k}\} \\ I \cup I' = \{1, \ldots, n\}}} P(A_{(i_1)} = 1, \ldots, A_{(i_k)} = 1, A_{(i'_1)} = 0, \ldots, A_{(i'_{n-k})} = 0) \qquad (2.22)$$

Eqn. (2.22) is a precise characterization of $P(A^n = k)$. Informally, we say that $P(A^n = k)$ determines the probability that event A occured k times after n repetitions of the experiment. The probability $P(A^n = k)$ meets the notion of binomial distribution that is defined as usual in Def. 2.10.

Definition 2.10 (Binomial Distribution)   Given a Bernoulli experiment with success probability p, a number $n \in \mathbb{N}$ of experiment repetitions and a number of successes $0 \leqslant k \leqslant n$. The *binomial distribution* w.r.t. to n and p, denoted by $\mathfrak{B}_{n,p}$, determines the probability of k successes after n experiment repetitions as follows:

$$\mathfrak{B}_{n,p}(k) = \binom{n}{k} p^k (1 - p)^{n-k} \qquad (2.23)$$

Now, given an i.i.d. sequence of random variables $(A_{(i)})_{i \in \mathbb{N}}$ we have that

$$P(A^n = k) = \mathfrak{B}_{n,P(A)}(k) \qquad (2.24)$$

Once more Note that $P(A) = P(A_{(1)}) = P(A_{(i)})$ in Eqn. (2.24). Also Note that $P(A^n = k)$ equals $P(\overline{A^n} = k/n)$. To model a series of simultaneously observed events, we can use an i.i.d. sequence of multivariate characteristic random variables $(\langle A_1, \ldots, A_m \rangle_{(i)})_{i \in \mathbb{N}}$; compare with Sect. 2.1.2. With multivariate characteristic random variable we mean that each of the marginal random variables A_j from $\langle A_1, \ldots, A_m \rangle$ is a characteristic random variable. As in the case of plain characteristic random variables above, we use A_j to denote the first random variable $A_{j(1)}$ for each marginal index $1 \leqslant j \leqslant m$. Again, we use $A_{j(i)}$ to denote the event $A_{j(i)} = 1$. Consequently, we use A_j to denote the event $A_{j(1)} = 1$. Furhermore, consequently, we use A_j^n to denote $A_{j(1)} + \cdots + A_{j(n)}$ and $\overline{A_j^n}$ to denote $\overline{A_{j(1)} + \cdots + A_{j(n)}}$. Now, we can say that $P(A_1^n = k_1, \ldots, A_m^n = k_m)$ models that each event A_j occured k_j times after n repetitions of the experiment. In general, it is not so convenient any more to determine the value of $P(A_1^n = k_1, \ldots, A_m^n = k_m)$ as it was in case of the plain characteristic random variable $P(A^n)$. However, in the special case that $A_1, \ldots, A_m$ form a partition, we have that $P(A_1^n = k_1, \ldots, A_m^n = k_m)$ meets the notion of multinomial distribution, which again provides a convenient combinatorial solution.

Definition 2.11 (Multinomial Distribution) Given an experiment with m mutually exclusive success categories and success probabilities $p_1, \ldots, p_m$ (i.e., such that $p_1 + \cdots + p_m = 1$), a number $n \in \mathbb{N}$ of repetitions and numbers of successes $k_1, \ldots, k_m$ for each category such that $k_1 + \cdots + k_m = n$. The *multinomial distribution* w.r.t. to n and $p_1, \ldots, p_m$, denoted by $\mathfrak{M}_{n,p_1,\ldots,p_m}$ determines the probability of k_j successes in all of the success categories j after n experiment repetitions as follows:

$$\mathfrak{M}_{n,p_1,\ldots,p_m}(k_1, \ldots, k_m) = \frac{n!}{k_1! \cdots k_m!} p_1 \cdots p_m \tag{2.25}$$

The multinomial distribution is a generalization of the binomial distribution, i.e., we have that $\mathfrak{B}_{n,p}(k)$ equals $\mathfrak{M}_{n,p,(1-p)}(k, n-k)$. Now, given an i.i.d.sequence of multivariate characteristic random variables $(\langle A_1, \ldots, A_m \rangle_{(i)})_{i \in \mathbb{N}}$ such that $A_1, \ldots, A_m$ form a partition, a number n and numbers $k_1, \ldots, k_m$ such that $k_1 + \cdots + k_m = n$ we have that

$$P(A_1^n = k_1, \ldots, A_m^n = k_m) = \mathfrak{M}_{n,P(A_1),\ldots,P(A_m)}(k_1, \ldots, k_m) \tag{2.26}$$

Again, note that $P(A_1^n = k_1, \ldots, A_m^n = k_m)$ equals $P(\overline{A_1^n} = k_1/n, \ldots, \overline{A_m^n} = k_m/n)$. The probability $P(A_1^n = k_1, \ldots, A_m^n = k_m)$ forms a straightforward, regular case, i.e., it shows an equal number of n repetitions for all involved sums. Of course, we can have arbitrary specifications $P(A_1^{n_1} = k_1, \ldots, A_m^{n_m} = k_m)$ for different values $n_1, \ldots, n_m$.

2.1.5 Aggregate Conditional Expected Values

The properties of conditional expected values are crucial in argumentations about F.P. conditionalizations. We walk through the most important properties in this section. Of course, the laws considered in this section also all apply in the unconditional case. Conditional expected values are additive, i.e., given two real-valued random variables $X : \Omega \longrightarrow \mathbb{R}$ and $Y : \Omega \longrightarrow \mathbb{R}$ and an event $C \subseteq \Omega$ we have the following:

$$E(X + Y \mid C) = E(X \mid C) + E(Y \mid C) \tag{2.27}$$

Actually, we have linearity for conditional expected values, i.e., we have the following for real numbers a and b:

$$E(a \cdot X + b \cdot Y \mid C) = a \cdot E(Y \mid C) + b \cdot E(X \mid C) \tag{2.28}$$

We are particularly interested in expected values of sequences of i.i.d. random variables. We are interested in various aggregates here. For the rest of the section, we deal with a sequence of i.i.d. random variables $X_1, \ldots, X_n$ and an event C such that $E(X_i|C)$ are equal for all $1 \leqslant i \leqslant n$. First, we have that the expected value of their sum X^n, conditional on C, equals n-times the expected value of each of the random variables X_i, conditional on C:

$$\forall 1 \leqslant i \leqslant n. E(X_i|C) = E(X|C) \implies \boxed{E(X^n \,|\, C) = n \cdot E(X \,|\, C)} \tag{2.29}$$

As usual, $E(X \,|\, C)$ serves as representative for all equal expectations $E(X_i \,|\, C)$ in the conclusion of Eqn. (2.29). Furthermore, we have that $\overline{X^n} = 1/n \cdot X^n$. Therefore, we have that the expected average of $X_1, \ldots, X_n$ conditional on C equals the expected value of each of the random variables X_i, conditional on C:

$$\forall 1 \leqslant i \leqslant n. E(X_i|C) = E(X|C) \implies \boxed{E(\overline{X^n} \,|\, C) = E(X \,|\, C)} \tag{2.30}$$

Next, let us furthermore assume that $X_1, \ldots, X_n$ are characteristic random variables. Now, we have that $E(X_i) = P(X)$ for all X_i in $X_1, \ldots, X_n$ and we also have that

$$X : \Omega \to \{0,1\}, \forall 1 \leqslant i \leqslant n. P(X_i|C) = P(X|C) \implies \boxed{\forall 1 \leqslant i \leqslant n. E(X_i|C) = P(X|C)} \tag{2.31}$$

In particular, we know that the expected average of characteristic random variables $X_1, \ldots, X_n$ conditional on C equals the probability of each random variable X_i conditional on C:

$$X : \Omega \to \{0,1\}, \forall 1 \leqslant i \leqslant n. P(X_i|C) = P(X|C) \implies \boxed{E(\overline{X^n} \,|\, C) = P(X \,|\, C)} \tag{2.32}$$

It is Eqn. (2.32) that is particularly important in the motivation and derivation of our definition of F.P. conditionalization later. The above side condition that $E(X_i|C) = E(X|C)$ for all $1 \leqslant i \leqslant n$ and its instance $P(X_i|C) = P(X|C)$ in case of characteristic random variables are essential for Eqns. (2.29) through (2.32), i.e., it is not sufficient that $X_1, \ldots, X_n$ are identically distributed as in the unconditional case, where we have that $E(X_i) = E(X)$ automatically holds. That is why Lemma C.3 is important for upcoming proofs, as it establishes a respective property for F.P. conditionalizations.

2.2 Defining Frequentist Partial Conditionalization

In the context of F.P. conditionalization, we want to denote a pair consisting of an event $B \subseteq \Omega$ and a probability $b \in [0,1]$ as $B \equiv b$ that we call frequency specification or also probability specification. We can read $B \equiv b$ as, e.g., B's probability is updated by b, B's probability becomes b, or simply, although not completely precise, B is updated by b or B becomes b. Given a list of frequency specifications, the idea is to consider a repeated experiment, which has an appropriate length to show outcomes that adhere to the specifications, i.e., an experiment in which the frequencies of the frequency specification can be realized for the respective events. Let us explain this in more depth. Let us assume that we have a sequence of i.i.d. characteristic random variables $(B_i)_{i \in \mathbb{N}}$ that models the repetition of B. Furthermore, let us assume that we have a single frequency specification $B \equiv b$ that has the irreducible fraction x/y as its representation. Now, a repeated experiment of y repetitions offers the possibility, that x/y of all single experiments end up in the event B. Also, any whole-number multiple $k \cdot y$ of repetitions offers this possibility. In all other cases of repetitions, it is not possible to have an outcome, in which x/y of the observed events are B. Now, let us assume that we have a whole number multiply n of y. Now, the fact that a fraction of b of the single experiments realize in the n-times repeated experiment, can be denoted via the averaged sum of observations of B as $\overline{B^n} = b$. This consideration can be generalized, in a straightforward manner, to the general case of a list of frequency specifications $B_1 \equiv b_1$ through $B_m \equiv b_m$. Let us come up with the definition of the so-called bounded F.P. conditionalization in Def. 2.12.

Definition 2.12 (Bounded F.P. Conditionalization)   Given an i.i.d. sequence of multivariate characteristic random variables $(\langle A, B_1, \ldots, B_m \rangle_{(j)})_{j \in \mathbb{N}}$, a list of rational numbers $b_1, \ldots, b_m$ and a bound $n \in \mathbb{N}$ such that $0 \leqslant b_i \leqslant 1$ and $nb_i \in \mathbb{N}$ for all b_i in $b_1, \ldots, b_m$. We define the *probability of A conditional on $B_1 \equiv b_1$ through $B_m \equiv b_m$ bounded by n*, which is denoted by $\mathsf{P}^n(A \mid B_1 \equiv b_1, \ldots, B_m \equiv b_m)$, as follows:

$$\mathsf{P}^n(A \mid B_1 \equiv b_1, \ldots, B_m \equiv b_m) = \mathsf{E}(\overline{A^n} \mid \overline{B_1^n} = b_1, \ldots, \overline{B_m^n} = b_m)$$

Let us come back to the intention behind bounded F.P. conditionalization. The idea is to repeat the considered experiment sufficiently often or to say it better, appropriately often, so that we can observe relative frequencies for all of the involved events B_i from the specified frequency specifications. Now, given such n-times repeated experiment, we would like to call this repeated experiment also an F.P. conditionalization testbed of length n, or testbed of length n for short. Now, we are interested in the outcomes that actually realize all of the specified frequencies. We keep them and throw away all other outcomes. With respect to the kept outcomes we are seeking for the relative frequencies that we can expect for the target event A in *and* relatively to these outcomes. And actually, this is what is achieved by the conditional expected value in Def. 2.12.

We have that an event $(\overline{B^n} = b)$ equals the event $(B^n = bn)$. We will use this fact throughout the book on many occasion without further mentioning. The classical conditional probability $\mathsf{P}(A|B) = \mathsf{P}(AB)/\mathsf{P}(B)$ is defined only in case $\mathsf{P}(B) \neq 0$. Similarly, the bounded F.P. conditionalization in Def. 2.12 is defined only in case $\mathsf{P}(\overline{B_1^n} = b_1, \ldots, \overline{B_m^n} = b_m) \neq 0$ and so it will be with F.P. conditionalization in general in Def. 2.14. We have chosen not to mention this side condition explicitly in definitions. Similarly, in Lemmas and proofs we assume that we are working with defined F.P. conditionalizations and that this side condition implicitly holds. Let us have a closer look on the violation of this side condition. For example, we know that $\mathsf{P}(\overline{B_1^n} = b_1, \ldots, \overline{B_m^n} = b_m)$ becomes zero whenever there is some $\overline{B_i^n} = b_i$ such that

$P(B_i) = 0$ but $b_i \neq 0$. As another example, $P(\overline{B_1}^n = b_1, \ldots, \overline{B_m}^n = b_m)$ becomes zero if the events $B_1, \ldots, B_m$ form a partition but $b_1 + \cdots + b_m \neq 1$.

The definition of bounded F.P. conditionalization in Def. 2.12 directly translates our intuition about F.P. conditionalization into a frequentist semantics. Fortunately, due to the linearity of expected values, it can be simplified leading to a more compact version of bounded F.P. conditionalization that comes as a conditional probability of the target event. We give this characterization of bounded F.P. conditionalization in Lemma 2.13.

Lemma 2.13 (Compact Bounded F.P. Conditionalization) *Given an F.P. conditionalization* $P^n(A \mid B_1 \equiv b_1, \ldots, B_m \equiv b_m)$ *we have that the following holds:*

$$P^n(A \mid B_1 \equiv b_1, \ldots, B_m \equiv b_m) = P(A \mid \overline{B_1}^n = b_1, \ldots, \overline{B_m}^n = b_m) \qquad (2.33)$$

Proof. Due to the definition of bounded F.P. conditionalization in Def. 2.12 we have that $P^n(A \mid B_1 \equiv b_1, \ldots, B_m \equiv b_m)$ equals

$$E(\overline{A}^n \mid \overline{B_1}^n = b_1, \ldots, \overline{B_m}^n = b_m) \qquad (2.34)$$

Now, the technically hard part is to proof that $P(A_{(i)} \mid \overline{B_1}^n = b_1, \ldots, \overline{B_m}^n = b_m)$ is the same for all $1 \leqslant i \leqslant n$. Fortunately, this is achieved by Lemma C.3. Based on that, we can apply Eqn. (2.32) to Eqn. (2.34) so that it equals $P(A \mid \overline{B_1}^n = b_1, \ldots, \overline{B_m}^n = b_m)$. $\square$

In case of a single frequency specification $B \equiv b$, we will see that the bounded F.P. conditionalizations $P^n(A \mid B \equiv b)$ are equal for all n. Therefore, we could take the value of any of these bounded F.P. conditionalizations as our definition of general, i.e., unbounded F.P. conditionalization $P(A \mid B \equiv b)$. However, in case of lists of frequency specifications $B_1 \equiv b_1, \ldots, B_m \equiv b_m$ things change. Here, $P^n(A \mid B_1 \equiv b_1, \ldots, B_m \equiv b_m)$ is, in general, different from $P^k(A \mid B_1 \equiv b_1, \ldots, B_m \equiv b_m)$ for all $k \neq n$. In general, we need to define $P(A \mid B_1 \equiv b_1, \ldots, B_m \equiv b_m)$ as the limit of $P^n(A \mid B_1 \equiv b_1, \ldots, B_m \equiv b_m)$ w.r.t. n growing infinitely in all appropriate bounds. We come up with the defintion in Def. 2.14. This time, we express the appropriate bounds on the basis of the least common denominator of all involved frequencies b_i, that we denote by $lcd(b_1, \ldots, b_m)$.

Definition 2.14 (F.P. Conditionalization)   Given an i.i.d. sequence of multivariate characteristic random variables $(\langle A, B_1, \ldots, B_m \rangle_{(j)})_{j \in \mathbb{N}}$ and a list of rational numbers $b = b_1, \ldots, b_m$ such that $0 \leqslant b_i \leqslant 1$ for all b_i in b. We define the *probability of A conditional on* $B_1 \equiv b_1$ *through* $B_m \equiv b_m$, denoted by $P(A \mid B_1 \equiv b_1, \ldots, B_m \equiv b_m)$, as follows:

$$P(A \mid B_1 \equiv b_1, \ldots, B_m \equiv b_m) = \lim_{k \to \infty} P^{k \cdot lcd(b)}(A \mid B_1 \equiv b_1, \ldots, B_m \equiv b_m) \qquad (2.35)$$

Our notion of F.P. conditionalization in Def. 2.14 is only defined for those cases in which the limit in Eqn. (2.35) exists. In some important cases we will see that the limit exists, simply because the bounded F.P. conditionalization is equal for all appropriate bounds in these cases. This is so in the cases of Jeffrey conditionalization and in the case of independent condition variables. The bounded F.P. conditionalizations $P^n(A \mid B_1 \equiv b_1, \ldots, B_m \equiv b_m)$ are the approximations to the F.P. conditionalization $P(A \mid B_1 \equiv b_1, \ldots, B_m \equiv b_m)$. In general, there is no finite testbed to yield the value of an F.P. conditionalization. An F.P. conditionalization is a transcendent object, i.e., we cannot think of its value in terms of a conventional executable experiment any more.

We call a tuple of the form $B \equiv b$ a frequency specification or probability specification. Similarly, we also talk about events of the form $B^n = b$ as frequency specifications and probability specifications. A frequency specification $B^n = b$ is, in a sense, an instance of the frequency specification $B \equiv b$. We call events of the form $(B_1^{n_1} = b_1, \ldots, B_m^{n_m} = b_m)$ F.P. test, F.P. testbeds, F.P. test specification and the like. And we do so for the more general events $(A, B_1^{n_1} = b_1, \ldots, B_m^{n_m} = b_m)$, i.e., we want to speak about an event as an F.P. test whenever some frequency specification is involved. When working with F.P. conditionalization you might find it useful to have also some special syntax for F.P. test probabilities $P(B_1^n = b_1, \ldots, B_m^n = b_m)$ to hand. It is straightforward to use $P^n(B_1 \equiv b_1, \ldots, B_m \equiv b_m)$ for such probabilities, however, in this book we avoid to use such further special syntax.

2.3 Projective F.P. Conditionalizations

Let us have a look at an important kind of F.P. conditionalization, which is particularly intuitive. If we assign some new *a postiori* probabilities to some events, we expect that each of these events actually has the *a postiori* probability that has been assigned to it after update, i.e., conditional on all probability assignments. And actually we have that, see Lemma 2.15. In such case, we say that the F.P. conditionalization is projective in a single of its condition events or a projective F.P. conditionalization for short.

Lemma 2.15 (Projective F.P. Conditionalizations) *Given a collection of probability specifications $B_1 \equiv b_1, \ldots, B_m \equiv b_m$ we have the following for each $1 \leqslant i \leqslant m$:*

$$P(B_i \mid B_1 \equiv b_1, \ldots, B_m \equiv b_m) = b_i \tag{2.36}$$

Proof. To see the correctness of Eqn. (2.36) it suffices to show it for all its finite approximations, i.e., that $P^n(B_i \mid B_1 \equiv b_1, \ldots, B_m \equiv b_m) = b_i$ for all appropriate bounds n. Due to the definition of F.P. conditionalization Def. 2.12 we have that $P^n(B_i \mid B_1 \equiv b_1, \ldots, B_m \equiv b_m)$ equals

$$E(\overline{B_i^n} \mid \overline{B_1^n} = b_1, \ldots, \overline{B_i^n} = b_i, \ldots, \overline{B_m^n} = b_m) \tag{2.37}$$

According to the definition of conditional expected values we have that Eqn. (2.37) equals

$$\sum_{k=0}^{n} (k/n) \cdot \frac{\mathsf{P}(\overline{B_i}^n = (k/n), \overline{B_1}^n = b_1, \dots, \overline{B_i}^n = b_i, \dots, \overline{B_m}^n = b_m)}{\mathsf{P}(\overline{B_1}^n = b_1, \dots, \overline{B_m}^n = b_m)} \tag{2.38}$$

Note that the sum in Eqn. (2.38) ranges over all values (k/n), in which the random variable $\overline{B_i}^n$ can potentially realize with a probability different from zero, i.e., k/n ranges over all possible relative occurences of B_i after n repetitions. Now, both the frequency specification $\overline{B_i}^n = (k/n)$ and the frequency specification $\overline{B_i}^n = b_i$ occur in the numerator event. So, if $k/n \neq b_i$ we have that the event $(\overline{B^n} = (k/n), \overline{B^n} = b_i)$ is empty and consequently the whole numerator event has a probability of zero. Otherwise, in case $k/n = b_i$ we have that the numerator event and the denumerator event are equal so that Eqn. (2.38) has the final value b_i.   □

Due to Lemma 2.15 we have that each single condition event B_i of a an F.P. conditionalization $\mathsf{P}^n(B_i \mid B_1 \equiv b_1, \dots, B_m \equiv b_m)$ actually equals its *a posteriori* probability b_i under this conditionalization. This is natural. However, things change, if we consider combinations of condition events $B_{i_1}, \dots, B_{i_k}$. In the special case that the condition events $B_1, \dots, B_m$ are themselves also mutually independent, we will see later that $\mathsf{P}^n(B_{i_1}, \dots, B_{i_k} \mid B_1 \equiv b_1, \dots, B_m \equiv b_m)$ equals the product $b_{i_1} \cdots b_{i_k}$ of the involved *a posteriori* probabilities. However, in general, such F.P. conditionalization is as hard to determine as any other F.P. conditionalization $\mathsf{P}^n(A \mid B_1 \equiv b_1, \dots, B_m \equiv b_m)$. For example, let us consider the case of only two condition events $\mathsf{P}^n(B_1 B_2 \mid B_1 \equiv b_1, B_1 \equiv b_2)$. Following the lines of the above argumentation we have that this equals $\mathsf{E}(\overline{B_1 B_2}^n \mid \overline{B_1}^n = b_1, \overline{B_2}^n = b_2)$ which equals

$$\sum_{k=0}^{n} (k/n) \cdot \frac{\mathsf{P}(\overline{B_1 B_2}^n = (k/n), \overline{B_1}^n = b_1, \overline{B_2}^n = b_2)}{\mathsf{P}(\overline{B_1}^n = b_1, \overline{B_2}^n = b_2)} \tag{2.39}$$

Now, we can see that in general there might be more than one (k/n) such that the conjunction of the events $(\overline{B_1 B_2}^n = (k/n))$, $(\overline{B_1}^n = b_1)$ and $(\overline{B_2}^n = b_2)$ is not empty. As a result we cannot proceed further with simplifying Eqn. (2.39) as we did in case of $\mathsf{P}(B_i \mid B_1 \equiv b_1, \dots, B_m \equiv b_m)$. Actually, we know the bounds in which (k/n) might range without making the numerator event in Eqn. (2.39) necessarily empty; compare with Sect. 4.1.2.

2.4 Cutting Repetitions

Shortening and cutting repetitions in frequency specifications is an important two-step pattern that is useful on many occasions. First, sums of random variables are shortened, while frequencies are adjusted. In the second step, the mutual independence between random variables is exploited to cut of a repetition. There might be a third step, in which the shortened repetitions are shifted to the first position; however, this can usually be considered a merely presentational issue.

Given a sequence of i.i.d. characteristic random variables $(B_i)_{i \in \mathbb{N}}$. Let us consider the following event for some $1 \leqslant k \leqslant n$:

$$(B, B^n = k) \tag{2.40}$$

We can rewrite Eqn. (2.40) as follows

$$(B_{(1)}, B_{(1)} + \ldots + B_{(n)} = k) \tag{2.41}$$

Now, in case of Eqn. (2.41) we know that event B has occurred k times after n repetitions; compare also with Eqn. (2.22). However, we furthermore know, that B has occurred upon the first repetition. Therefore we know that event B must have occurred $k - 1$ times in the last $n - 1$ repetitions, i.e., in the repetitions 2 through n. Therefore we know that Eqn. (2.41) equals

$$(B_{(1)}, B_{(2)} + \ldots + B_{(n)} = k - 1) \tag{2.42}$$

Concerning the step from Eqn. (2.40) to Eqn. (2.42) we say that we have shortened the sum $B_{(1)} + \ldots + B_{(n)}$ to $B_{(2)} + \ldots + B_{(n)}$ and adjusted the frequency k to $k - 1$. With respect to Eqn. (2.40) we have established the side-condition $k \neq 0$. We did so, because in case $k = 0$ we have that Eqn. (2.40) imposes a conflict. In case $B^n = 0$ we have that B does not occur at all in n repetitions and therefore it can also not occur upon the first repetition. We have that B is disjoint from $B^n = 0$ in $(B, B^n = 0)$. Therefore we have that $(B, B^n = 0)$ is empty, i.e.:

$$(B, B^n = 0) = \emptyset \tag{2.43}$$

Now, let us consider the following event for some $0 \leqslant k < n$:

$$(\overline{B}, B^n = k) \tag{2.44}$$

Again, we can rewrite Eqn. (2.44), this time yielding

$$(\overline{B_{(1)}}, B_{(1)} + \ldots + B_{(n)} = k) \tag{2.45}$$

We can see that in case of Eqn. (2.45) all k occurrences must occur during the last $n - 1$ repetitions, because B has not occurred upon the first repetition. Therefore, we have that Eqn. (2.45) equals

$$(\overline{B_{(1)}}, B_{(2)} + \ldots + B_{(n)} = k) \tag{2.46}$$

Furthermore, we know that in case $k = n$ the event Eqn. (2.44) must be empty. In case $B^n = n$ we have that B must occur upon each of the n repetitions and therefore it must also occur upon the first repetition. We have that $\overline{B}$ is disjoint from $B^n = n$ and therefore we have that

$$(\overline{B}, B^n = n) = \emptyset \tag{2.47}$$

In Lemma 2.16 we summarize the results found so far in this section for further reference. Here, we assume that a sum $B_{(i)} + \ldots + B_{(i')}$ with $i > i'$ stands for the sum of length zero B^0; compare with Eqn. (2.18). This is relevant to Eqns. (2.48) and (2.49) where $B_{(2)} + \ldots + B_{(n)}$ turns into $B_{(2)} + \ldots + B_{(1)}$ in case $n = 1$.

Lemma 2.16 (Shortening and Adjusting) *Given a sequence of i.i.d. characteristic random variables* $(B_i)_{i \in \mathbb{N}}$, *a number of repetitions* $n \in \mathbb{N}$ *and (absolute) frequencies* $1 \leqslant k \leqslant n$ *and* $0 \leqslant k' < n$ *we have the following:*

$$(B, B^n = k) = (B, B_{(2)} + \ldots + B_{(n)} = k - 1) \tag{2.48}$$

$$(\overline{B}, B^n = k') = (\overline{B}, B_{(2)} + \ldots + B_{(n)} = k') \tag{2.49}$$

$$(B, B^n = 0) = \emptyset \tag{2.50}$$

$$(\overline{B}, B^n = n) = \emptyset \tag{2.51}$$

$$(B, B^1 = 1) = B \tag{2.52}$$

$$(\overline{B}, B^1 = 0) = \overline{B} \tag{2.53}$$

Proof. With respect to Eqns. (2.48) through (2.51) compare with Eqns. (2.40) through (2.47), where we once more note that $B = B_{(1)}$. Eqns. (2.52) and (2.53) trivially hold due to $(B^1 = 1) = B$ and $(B^1 = 0) = \overline{B}$, actually, they are redundant to Eqns. (2.48) and (2.49). $\qquad\qquad\qquad\qquad\qquad\qquad\qquad\qquad\qquad\qquad\qquad\qquad \square$

Now, due to Corollary 2.9 we can apply Lemma 2.16 to the sequences of marginals of multivariate random variables. Note that the equations in Lemma 2.16 are set equalities of the form $A = A'$ and as such are stronger than respective equations of the form $\mathsf{P}(A) = \mathsf{P}(A')$. In their form $A = A'$ they can be directly exploited, without any further technical effort, for transforming frequency specifications in marginal sequences, i.e., also in the context of other marginal sequences.

The equations in Lemma 2.16 can all be rewritten in terms of averages and relative frequencies, e.g., turning Eqns. (2.48) and (2.49) into

$$(B, \overline{B^n} = b) = (B, \overline{B_{(2)} + \ldots + B_{(n)}} = \frac{bn - 1}{n - 1}) \tag{2.54}$$

$$(\overline{B}, \overline{B^n} = b) = (\overline{B}, \overline{B_{(2)} + \ldots + B_{(n)}} = \frac{bn}{n - 1}) \tag{2.55}$$

Again, in Eqns. (2.54) and (2.55) we say that the frequency b is adjusted to $(bn - 1)/(n - 1)$ resp. $(bn)/(n - 1)$. Working with relative frequencies can be considered more intuitive because relative frequencies are probability values. However, working with absolute frequencies is usually more convenient for us in proofs, and therefore we mostly stay with the form of Lemma 2.16.

We can exploit the fact that $(B_i)_{i \in \mathbb{N}}$ is a sequence of i.i.d. random variables further. Let us consider the following probability:

$$\mathsf{P}(B, B^n = k) \tag{2.56}$$

Due to Lemma 2.16 we know that 2.56 equals

$$\mathsf{P}(B_{(1)}, B_{(2)} + \ldots + B_{(n)} = k - 1) \tag{2.57}$$

Due the mutual independence of $B_{(1)}, \ldots, B_{(n)}$ we immediately have that 2.57 equals

$$P(B_{(1)}) \cdot P(B_{(2)} + \ldots + B_{(n)} = k - 1) \tag{2.58}$$

Now, due to Eqn. (2.16) we have that Eqn. (2.58) equals

$$P(B_{(1)}) \cdot P(B_{(1)} + \ldots + B_{(n-1)} = k - 1) \tag{2.59}$$

Finally, we can rewrite Eqn. (2.59) so that we arrive at the following equation:

$$P(B, B^n = k) = P(B) \cdot P(B^{n-1} = k - 1) \tag{2.60}$$

Similarly, we have that $P(\overline{B}, B^n = k)$ equals $P(\overline{B}) \cdot P(B^{n-1} = k)$. The mutual independence of $B_{(1)}, \ldots, B_{(n)}$ is necessary to transform a probability of the form $P(B_{(1)}, B_{(2)} + \ldots + B_{(n)} = k)$ further.

2.5 Conditional Events

In Sect. 2.2 we have established a frequentist semantics for partial conditional probability. But how about the frequentist semantics of classical conditional probabilities? And why is that a question at all? These issues are addressed in this section. We will see, how the usual definition of conditional probabilities is met by a translation of the intuitive *operational semantics* of conditional probabilities into concrete, formal probabilities. Then, we will use the formalized operational semantics to turn the strong law of large numbers into a version that works for conditional probabilities, this way reinforcing the notion of conditional probability and its classical definition.

The conditional probability $P(A|B)$ is defined as $P(AB)/P(B)$; compare with Eqn. (1.1). Informally, it can be explained as the frequency of A in all of those experiments that succeeded in B; compare with the explanation on pp. 3. But how does this meet the definition of $P(A|B)$ as $P(AB)/P(B)$? And how does this formally show in a corresponding law of large numbers? To answer these questions, let us first turn the above informal description of conditional probability into a more precise, but yet still informal, operational semantics. We can think of the experiment e' modeled by a conditional probability P_B as two-staged. An agent a_1 triggers the experiment e'. On behalf of this an agent a_2 repeatedly executes the experiment e modeled by P until event B eventually occurs. Then, a_2 yields back the outcome of the last repetition of e – as result of e' – to agent a_1. Agent a_2 throws away experiments; to agent a_1 it appears as if events that show $\overline{B}$ never occur. Note that the composed experiment e' might be considered as not always terminating. However, it is *(almost) sure* that e' terminates; compare this also with the notion of termination degree as defined and discussed in [44].

With these considerations in mind let us consider a repeated experiment with the i.i.d. sequence $(\langle A_{(i)}, B_{(i)} \rangle)_{i \in \mathbb{N}}$. Now we define the event A_B that we might want to call a conditional event or just A given B for short:

$$A_B = \bigcup_{\eta \in \mathbb{N}} (\overline{B}_{(1)}, \ldots, \overline{B}_{(\eta-1)}, B_{(\eta)}, A_{(\eta)}) \tag{2.61}$$

Actually, we will show that $P(A_B)$ equals $P(AB)/P(B)$, i.e.,

$$\boxed{P_B(A) = P(A_B)} \tag{2.62}$$

See, how A_B turns our operational semantics from above into a concrete event. A_B consists of exactly those outcomes that show success in A when they have succeeded in B for the first time, i.e., a repeated experiment proceeds with failing in B until it succeeds in B in some repetition η and then belongs to A_B when it also succeeded in A in this repetition η. This is what our operational semantics is about. We will now proof that $P_B(A) = P(A_B)$. We do this in a generalized form. We step from first occurrences of B to events conditional on the n-th occurrence of B. The probability of the resulting $A_{B(n)}$ is again equal to $P_B(A)$. It turns out that the sequence $(A_{B(i)})_{i \in \mathbb{N}}$ is i.i.d., which explains why it is important to deal with the general concept, because on the basis of $(A_{B(i)})_{i \in \mathbb{N}}$ we will see how the strong law of large numbers works for conditionals.

Definition 2.17 (Conditional Event) Given a sequence of i.i.d. multivariate characteristic random variables $(\langle A_{(i)}, B_{(i)} \rangle)_{i \in \mathbb{N}}$, we define the *conditional event* $A_{B(n)}$ for each **n**-th first occurrence of B, also called *hitting time* **n** of B, where we use $\#I$ to denote the size of set I, as follows:

$$A_{B(\mathbf{n})} = \bigcup_{\substack{\eta \in \mathbb{N} \\ I \subseteq \{1, \ldots, \eta-1\} \\ I' = \{1, \ldots, \eta-1\} \setminus I \\ \#I = (\mathbf{n}-1)}} (\underset{i \in I}{\cap} B_{(i)}, \underset{i' \in I'}{\cap} \overline{B}_{(i')}, B_{(\eta)}, A_{(\eta)}) \tag{2.63}$$

Of course, we have that Eqn. (2.61) is an instance of Def. 2.17 with $A_B = A_{B(1)}$. We start with a technical helper Lemma 2.18 that we need in upcoming proofs. It expresses that it does not matter where to start the observation of an event $A_{B(n)}$. The lemma follows rather immediately from the i.i.d. property of the underlying random variables. Then we proof that $P(A_B) = P(A_{B(n)})$ for each hitting time n in Theorem 2.19. The proof idea is to decompose the probabilities $P(A_{B(n)})$ and this way to establish a linear equation system that determines the targeted probabilities as its unique solution. This technique is borrowed from Markov chains, where it is a basic pattern to analyze hitting and absorption probabilities; compare also with the notion of reachability in [44].

Lemma 2.18 (Start Times for Observing) *Given a sequence of i.i.d. multivariate characteristic random variables $(\langle A_{(i)}, B_{(i)} \rangle)_{i \in \mathbb{N}}$ and an offset $m \in \mathbb{N}$, we have the following for each hitting time **n**:*

$$P(A_{B(\mathbf{n})}) = \quad P\Big(\bigcup_{\substack{\eta \in \mathbb{N} \\ I \subseteq \{1+m,\ldots,\eta-1+m\} \\ I' = \{1+m,\ldots,\eta-1+m\}\setminus I \\ \#I = (\mathbf{n}-1)}} (\bigcap_{i\in I} B_{(i)}, \bigcap_{i'\in I'} \overline{B}_{(i')}, B_{(\eta+m)}, A_{(\eta+m)})\Big) \qquad (2.64)$$

Proof. It is possible to see that the events for all η in Eqn. (2.64) are all mutually disjoint. Therefore, we have that $P(A_{B(\mathbf{n})})$ equals

$$\sum_{\substack{\eta \in \mathbb{N} \\ I \subseteq \{1+m,\ldots,\eta-1+m\} \\ I' = \{1+m,\ldots,\eta-1+m\}\setminus I \\ \#I = (\mathbf{n}-1)}} P(\bigcap_{i\in I} B_{(i)}, \bigcap_{i'\in I'} \overline{B}_{(i')}, B_{(\eta+m)}, A_{(\eta+m)}) \qquad (2.65)$$

Due the lemma's premise we have that $(\langle A_{(i)}, B_{(i)}\rangle)_{i\in\mathbb{N}}$ is i.i.d.. Therefore, with setting $J = \{(i-m)|i \in I\}$ and $J' = \{(i'-m)|i' \in I'\}$ we have that each summand in Eqn. (2.65) equals

$$P(\cap_{j\in J} B_{(j)}, \cap_{j'\in J'} \overline{B}_{(j')}, B_{(\eta)}, A_{(\eta)}) \qquad (2.66)$$

Based on Eqn. (2.66) it follows immediately that Eqn. (2.65) equals $P(A_{B(\mathbf{n})})$. □

Theorem 2.19 (Probability of Conditional Events)   *Given a sequence of i.i.d. multivariate characteristic random variables* $(\langle A_{(i)}, B_{(i)}\rangle)_{i\in\mathbb{N}}$, *we have the following for each hitting time n:*

$$P_B(A) = P(A_{B(n)}) \qquad (2.67)$$

Proof. We distinguish two cases, i.e., the case that the hitting time n equals one and the case that it is greater than one. We start with the case $P(A_{B(1)})$. We can segment the probability $P(A_{B(1)})$ via $B_{(1)}$ so that it equals

$$P(B_{(1)}, A_{B(1)}) + P(\overline{B}_{(1)}, A_{B(1)}) \qquad (2.68)$$

Due to Def. 2.17 we have that $P(B_{(1)}, A_{B(1)})$ equals $P(B_{(1)}, A_{(1)})$, which is $P(AB)$. Therefore, we have that Eqn. (2.68) equals

$$P(AB) + P(\overline{B}_{(1)}, A_{B(1)}) \qquad (2.69)$$

Now, let us turn to the second summand in Eqn. (2.69). Due to Def. 2.17 we have that $P(\overline{B}_{(1)}, A_{B(1)})$ equals

$$P\Big(\overline{B}_{(1)}, \bigcup_{\eta\in\mathbb{N}} (\overline{B}_{(1)}, \ldots, \overline{B}_{(\eta-1)}, B_{(\eta)}, A_{(\eta)})\Big) \qquad (2.70)$$

Now, in presence of $\overline{B}_{(1)}$ we know that the first occurrence of B can be no earlier than after the 2nd repetition, which means that we can increase the range index η in Eqn. (2.70) by one, and similarly can increase the lower bound of $\overline{B}_{(1)}, \ldots, \overline{B}_{(\eta-1)}$ in Eqn. (2.70) by one without changing the overall event, resulting into

$$P\left(\overline{B}_{(1)}, \bigcup_{\eta \in \mathbb{N}} (\overline{B}_{(2)}, ..., \overline{B}_{(\eta)}, B_{(\eta+1)}, A_{(\eta+1)})\right) \tag{2.71}$$

Due to the premise that $(\langle A_{(i)}, B_{(i)} \rangle)_{i \in \mathbb{N}}$ is i.i.d. , we have that Eqn. (2.71) equals

$$P(\overline{B}_{(1)}) \cdot P\left(\bigcup_{\eta \in \mathbb{N}} (\overline{B}_{(2)}, ..., \overline{B}_{(\eta)}, B_{(\eta+1)}, A_{(\eta+1)})\right) \tag{2.72}$$

Now, due to Lemma 2.18 we have that Eqn. (2.72) equals $P(\overline{B}_{(1)}) \cdot P(A_{B(1)})$. Summarizing so far, we have that the following equation is valid:

$$P(A_{B(1)}) = P(AB) + P(\overline{B}) \cdot P(A_{B(1)}) \tag{2.73}$$

With analogue arguments it is possible to show that the following equation holds true for all $1 < i \leqslant n$:

$$P(A_{B(i)}) = P(B) \cdot P(A_{B(i-1)}) + P(\overline{B}) \cdot P(A_{B(i)}) \tag{2.74}$$

Now, Eqns. (2.73) and (2.74) form a finite linear equation system with n equations, where we take $P(A_{B(1)}), ..., P(A_{B(n)})$ as variables. The equation system determines $P(A_{B(i)}) = P(AB)/P(B)$ for all $1 \leqslant i \leqslant n$ as solution. To see this, we transform Eqn. (2.73) into

$$P(A_{B(1)})(1 - P(\overline{B})) = P(AB) \tag{2.75}$$

As $(1 - P(\overline{B}))$ equals $P(B)$ we have immediately that $P(A_{B(1)})$ equals $P(AB)/P(B)$. The rest of the equation system is solved by successive variable substitution. $\square$

With Theorem 2.19 we have that $P(A_B)$ yields an alternative definition of $P_B(A)$. We say that $P(A_B)$ is purely summative, which is explained by its infinite extension $P(AB) + P(\overline{B})(P(AB) + P(\overline{B})(P(AB) + P(\overline{B})(\cdots$

We proceed with the fact that the random variables $(A_{B(i)})_{i \in \mathbb{N}}$ are mutually independent in Lemma 2.20.

Lemma 2.20 (Mutual Independence of Conditional Events) *Given a sequence of i.i.d. multivariate characteristic random variables* $(\langle A_{(i)}, B_{(i)} \rangle)_{i \in \mathbb{N}}$, *and indices* $i_1, ..., i_n$ *we have the following:*

$$P(A_{B(i_1)}, ..., A_{B(i_n)}) = P(A_{B(i_1)}) \times \cdots \times P(A_{B(i_n)}) \tag{2.76}$$

Proof. (Sketch) With arguments analogue to those encountered in Theorem 2.19 it is possible to establish a finite linear equation system that determines, as part of its unique solution, the value of $P(A_{B(i_1)}, ..., A_{B(i_n)})$ as $P_B(A)^n$. With Theorem 2.19 we have that $P_B(A)^n$ equals $P(A_{B(i_1)}) \times \cdots \times P(A_{B(i_n)})$. $\square$

Corollary 2.21 (Conditional Events are I.I.D.) *Given a sequence of i.i.d. multivariate characteristic random variables* $(\langle A_{(i)}, B_{(i)} \rangle)_{i \in \mathbb{N}}$, *we have that the sequence of characteristic random variables* $(A_{B(i)})_{i \in \mathbb{N}}$ *is i.i.d.*

Proof. Immediate Corollary from Theorem 2.19 and Lemma 2.20.   □

Corollary 2.21 is a crucial result. With $(A_{B(i)})_{i\in\mathbb{N}}$ being i.i.d. we can exploit the law of large numbers to explain the behavior of conditional probabilities in the limit. We also say that $(A_{B(i)})_{i\in\mathbb{N}}$ is a conditional sequence of random variables. With respect to $(A_{(i)})_{i\in\mathbb{N}}$, the strong law of large numbers expresses that the limit of relative frequencies $\overline{A^n}$ meets the expected value of A in the following sense:

$$P(\lim_{n\longrightarrow\infty} \overline{A^n} = \mu_A) = 1 \qquad (2.77)$$

Remember that the event in Eqn. (2.77) stands for $((lim_{n\to\infty} \overline{A^n}) = \mu_A)$ and not for $(lim_{n\to\infty} (\overline{A^n} = \mu_A))$. The probability $P(lim_{n\to\infty} (\overline{A^n} = \mu_A))$ equals zero. It is the pointwise limit $(lim_{n\to\infty} \overline{A^n})(\omega)$ of all outcomes ω that makes the essence of the strong law of large numbers. Furthermore, we have that μ_A stands for the expected value $E(A)$. As we consider conditional probabilities $P(A|B)$ where all random variables in $(\langle A_{(i)}, B_{(i)} \rangle)_{i\in\mathbb{N}}$ are characteristic random variables, we have that μ_A equals $P(A)$. Now that $(A_{B(i)})_{i\in\mathbb{N}}$ is i.i.d. with respect to the measure P with $\mu_{A_B} = P_B(A)$, we can apply the law of large numbers yielding

$$P(\lim_{n\longrightarrow\infty} \overline{A^n_B} = P_B(A)) = 1 \qquad (2.78)$$

Equation (2.78) brings together our intuition of conditional probability, expressed by our operational semantics and formalized by the notion of conditional sequence of random variables $(A_{B(i)})_{i\in\mathbb{N}}$, with the classical definition of conditional probability P_B via the strong law of large numbers, this way reinforcing the definition of the probability measure P_B.

Chapter 3
F.P. Semantics of Jeffrey Conditionalization

In this chapter, we analyze F.P. conditionalization in case the condition events form a partition. Actually, this is the Jeffrey conditionalization case, as Jeffrey conditionalization is defined for exactly this special case. It turns out that F.P. conditionalization adheres to Jeffrey's rule in this case. You can see this result also from a different angle. It can be said, that F.P. conditionalization provides a frequentist semantics for Jeffrey conditionalization. In Sect. 3.1 we look at the basic case of F.P. conditionalization with respect to a single frequency specification. In Sect. 3.2 we treat the case of arbitrary partitions.

3.1 The Case of Basic Jeffrey Generalization

Given a single frequency specification $B \equiv b$ such that neither $P(B)$ nor $P(\overline{B})$ equals zero, the F.P. conditionalization $P(A \mid B \equiv b)$ turns out to be the weighted average of the conditional probabilities $P(A \mid B)$ and $P(A \mid \overline{B})$, weighted by the respective *a posteriori* frequencies b and $1 - b$ of B and $\overline{B}$. Here, b has been explicitly specified as frequency for the event B, whereas $1 - b$ is implicitly specified for the complement set $\overline{B}$. This weighted average of conditional probabilities is exactly what is proven next by Lemma 3.1. Actually, in the special case that $P(B) = 0$ and $b = 0$ we have that $P(A \mid B \equiv b)$ equals $P(A)$ and so it is in the dual special case that $P(B) = 1$ and $b = 1$. We have singled out these special cases into an own little Lemma 3.2, as the rule in Lemma 3.1 is the crucial one. The proofs of Lemmas 3.1 and Lemma 3.2 can be skipped without great loss by readers who are in a hurry, as the lemmas are just instances of Theorem 3.3 that deals with partitions of arbitrary length.

Lemma 3.1 (F.P. Conditionalization over a Single Condition) *Given an F.P. conditionalization* $P(A \mid B \equiv b)$ *for a single frequency specification* $B \equiv b$ *such that* $0 < P(B) < 1$ *we have the following:*

$$P(A \mid B \equiv b) = b \cdot P(A \mid B) + (1 - b) \cdot P(A \mid \overline{B}) \qquad (3.1)$$

© The Author(s) 2017

D. Draheim, *Generalized Jeffrey Conditionalization*, SpringerBriefs
in Computer Science, https://doi.org/10.1007/978-3-319-69868-7_3

Proof. The idea is to proof that Eqn. (3.1) holds for all of its approximations. Then Eqn. (3.1) follows immediately as a corollary. Due to the defintion of F.P. conditionalization Def. 2.14 we have that $(\langle A, B \rangle_{(i)})_{i \in \mathbb{N}}$ is a sequence of i.i.d. multivariate characteristic random variables. Now, we consider the bounded F.P. conditionalization $\mathsf{P}^n(A \mid B \equiv b)$ for any arbitrary but fixed $n \in \mathbb{N}$ such that $n \cdot b \in \mathbb{N}$, and will proof that the following holds:

$$\mathsf{P}^n(A \mid B \equiv b) = b \cdot \mathsf{P}(A \mid B) + (1 - b) \cdot \mathsf{P}(A \mid \overline{B}) \tag{3.2}$$

We start with the left-hand side of Eqn. (3.2). Due to Lemma. 2.13 we have that $\mathsf{P}^n(A \mid B \equiv b)$ equals

$$\mathsf{P}(A \mid B^n = bn) \tag{3.3}$$

Due to the definition of conditional probability we have that Eqn. (3.3) equals

$$\mathsf{P}(A, B^n = bn)/\mathsf{P}(B^n = bn) \tag{3.4}$$

The idea is to split the probability in Eqn. (3.4) into a sum by segmenting the event A via the event B and its complement $\overline{B}$. Next, we will shorten the number of repeated experiments by one and adjust the frequencies to the shorter lengths of repeated experiments. First, we have that Eqn. (3.4) equals

$$\frac{\mathsf{P}(AB, B^n = bn)}{\mathsf{P}(B^n = bn)} + \frac{\mathsf{P}(A\overline{B}, B^n = bn)}{\mathsf{P}(B^n = bn)} \tag{3.5}$$

We start with the first summand. We need to distinguish the cases $b = 0$ and $b \neq 0$. In case $b = 0$ we have, due to to Lemma 2.16, Eqn. (2.50), that $(AB, B^n = 0) = \emptyset$, hence the first summand equals zero and we are done. In case $b \neq 0$ we have, due to Eqn. (2.48), that the first summand equals

$$\frac{\mathsf{P}(AB, B_{(2)} + \cdots + B_{(n)} = bn - 1)}{\mathsf{P}(B^n = bn)} \tag{3.6}$$

Now, we can exploit the independence Lemma C.1 together with the fact that $(\langle A, B \rangle_{(i)})_{i \in \mathbb{N}}$ is i.i.d. to cut off $\mathsf{P}(AB)$ in Eqn. (3.6) yielding

$$\frac{\mathsf{P}(AB) \cdot \mathsf{P}(B_{(2)} + \cdots + B_{(n)} = bn - 1)}{\mathsf{P}(B^n = bn)} \tag{3.7}$$

Now, we can apply Eqn. (2.16) to Eqn. (3.7) resulting into

$$\mathsf{P}(AB) \cdot \mathsf{P}(B^{n-1} = bn - 1)/\mathsf{P}(B^n = bn) \tag{3.8}$$

Now, due to the premise that $\mathsf{P}(B) \neq 0$ we have that $\mathsf{P}(AB)$ equals $\mathsf{P}(A \mid B) \cdot \mathsf{P}(B)$ and therefore Eqn. (3.8) equals

$$\frac{\mathsf{P}(A \mid B) \cdot \mathsf{P}(B) \cdot \mathsf{P}(B^{n-1} = bn - 1)}{\mathsf{P}(B^n = bn)} \tag{3.9}$$

As the next crucial step, we have that B^n and B^{n-1} determine binomial distributions $\mathfrak{B}_{n,P(B)}$ and $\mathfrak{B}_{n-1,P(B)}$ so that we can resolve $P(B^{n-1} = bn-1)$ and $P(B^n = bn)$ combinatorically, yielding

$$P(A\,|\,B) \cdot P(B) \cdot \frac{\binom{n-1}{bn-1} \cdot P(B)^{bn-1} \cdot P(\overline{B})^{n-bn}}{\binom{n}{bn} \cdot P(B)^{bn} \cdot P(\overline{B})^{n-bn}} \tag{3.10}$$

As a next step, we can cancel all occurrences of $P(B)$ and $P(\overline{B})$ from Eqn. (3.10) and resolve the binomial coefficients resulting into

$$P(A\,|\,B) \cdot \frac{(n-1)!}{(bn-1)!(n-1-(bn-1))!} \,\Big/\, \frac{n!}{(bn)!(n-bn)!} \tag{3.11}$$

After resolving $(n-1)!$ to $n!/n$ and resolving $(bn-1)!$ to $(bn)!/(bn)$ we have that Eqn. (3.11) equals

$$P(A\,|\,B) \cdot \frac{n!\,bn}{n\,(bn)!(n-bn)!} \cdot \frac{(bn)!(n-bn)!}{n!} \tag{3.12}$$

Now, after a series of further cancelations we have that Eqn. (3.12) equals $b \cdot P(A\,|\,B)$ and we are done with the first summand in Eqn. (3.5). Now, let us turn to the second summand in Eqn. (3.5). Again, we need to distinguish two cases, i.e., $b = 1$ and $b \neq 1$ this time. In case $b = 1$ we have, due to Eqn. (2.51) that the second summand equals zero and we are done. In case $b \neq 1$ we have, due to Lemma 2.16, Eqn. (2.49), Lemma C.1, Eqn. (2.16), the premise that $P(B) \neq 1$ and the definition of conditional probability that the second summand equals

$$\frac{P(A\,|\,\overline{B}) \cdot P(\overline{B}) \cdot P(B^{n-1} = bn)}{P(B^n = bn)} \tag{3.13}$$

Now, we resolve $P(B^{n-1} = bn)$ and $P(B^n = bn)$ in Eqn. (3.13) combinatorially, yielding

$$P(A\,|\,\overline{B}) \cdot P(\overline{B}) \cdot \frac{\binom{n-1}{bn} \cdot P(B)^{bn} \cdot P(\overline{B})^{n-1-bn}}{\binom{n}{bn} \cdot P(B)^{bn} \cdot P(\overline{B})^{n-bn}} \tag{3.14}$$

Again, we can cancel all occurrences of $P(B)$ and $P(\overline{B})$ from Eqn. (3.14) which yields

$$P(A\,|\,\overline{B}) \cdot \frac{(n-1)!}{(bn)!(n-1-bn)!} \,\Big/\, \frac{n!}{(bn)!(n-bn)!} \tag{3.15}$$

Eqn. (3.15) can be transformed into

$$P(A\,|\,\overline{B}) \cdot \frac{n!\,(n-bn)}{n\,(bn)!(n-bn)!} \cdot \frac{(bn)!(n-bn)!}{n!} \tag{3.16}$$

Finally, after a series of further cancelations we have that Eqn. (3.16) equals $(1-b) \cdot P(A\,|\,\overline{B})$. $\qquad\square$

Lemma 3.2 (F.P. Conditionalization – Single Condition, Special Values) *Given an F.P. conditionalization* $P(A \mid B \equiv b)$ *for a single frequency specification* $B \equiv b$ *such that either* $P(B) = 0$ *and* $b = 0$ *or otherwise* $P(B) = 1$ *and* $b = 1$ *we have that*

$$P(A \mid B \equiv b) = P(A) \tag{3.17}$$

Proof. Let us assume that $P(B) = 0$. For each number n of repetitions we then have that $P(\overline{B^n} = 0) = 1$, see Eqn. (2.24). We therefore have that $P(A, \overline{B^n} = 0) = P(A)$; compare with Lemma C.5. Due Lemma 2.13 we have that $P^n(A \mid B \equiv b)$ equals $P(A, \overline{B^n} = 0)/P(\overline{B^n} = 0)$. Altogether, we have that $P^n(A \mid B \equiv b)$ equals $P(A)$. The dual case that $P(B) = 1$ and $b = 1$ holds analogously.   □

Given that $P(A|B \equiv b)$ equals $b \cdot P(A|B) + (1 - b) \cdot P(A|\overline{B})$ a series of intuitive examples are particularly easy to prove. For example, if we assign to a single event B its old probability $P(B)$ as its new probability, we have that the probabilities of other events A remain unchanged, i.e., we have that $P(A|B \equiv P(B))$ equals $P(B)$. To see this, first transform $P(A|B \equiv P(B))$ into $P(B) \cdot P(A|B) + P(\overline{B}) \cdot P(A|\overline{B})$. Now, we see that this equals $P(A, B) + P(A, \overline{B})$, which equals $P(A)$. We will be able to prove generalizations of $P(A|B \equiv P(B)) = P(B)$ in Lemma 4.7 and Lemma 4.8, where we show that $P(A|B_1 \equiv P(B_1), \ldots, B_m \equiv P(B_m))$ equals $P(A)$ for the cases that $B_1, \ldots, B_m$ form a partition or are mutually independent. Next, we see that updating the probability to 100% amounts to classical conditional probability, as $P(A|B \equiv 100\%)$ equals $100\% \cdot P(A|B) + 0\% \cdot P(A|B)$. Again, compare with the more general result in Lemma 4.6.

3.2 The Case of Jeffrey Conditionalization in General

A Jeffrey conditionalization $P(A \mid B_1 \equiv b_1, \ldots, B_m \equiv b_m)_J$ treats the special case that the events $B_1, \ldots, B_m$ form a partition of the outcome space. Henceforth, we just say that events form a partition if they form a partition of the outcome space.

Convention 3 (Partitions of the Outcome Space) If some events $B_1, \ldots, B_m$ are said to form a partition, this means that they form a partition of the outcome space.

Jeffrey conditionalization equals the weighted sum of the conditional probabilities $P(A \mid B_1), \ldots, P(A \mid B_m)$, weighted by the respective *a posteriori* probabilities $b_1, \ldots, b_m$; compare with Eqns. (1.3) and (1.6). In Theorem 3.3 we proof that in case of a partition $B_1, \ldots, B_m$ an F.P. conditionalization $P(A \mid B_1 \equiv b_1, \ldots, B_m \equiv b_m)$ equals Jeffrey conditionalization. We also say that with Theorem 3.3 we provide an F.P. semantics for Jeffrey conditionalization. Theorem 3.3 is a straightforward generalization of Lemma 3.1. It is just technically more complex.

Theorem 3.3 (F.P. Conditionalization over Partitions) *Given an F.P. conditionalization* $P(A \mid B_1 \equiv b_1, \ldots, B_m \equiv b_m)$ *such that the events* $B_1, \ldots, B_m$ *form a partition, and, furthermore, the frequencies* $b_1, \ldots, b_m$ *sum up to one, we have the following:*

$$P(A \mid B_1 \equiv b_1, \ldots, B_m \equiv b_m) = \sum_{\substack{1 \leqslant i \leqslant m \\ P(B_i) \neq 0}} b_i \cdot P(A \mid B_i) \tag{3.18}$$

Proof. We proof Eqn. (3.18) for all of its approximations. Due to Lemma. 2.13 we have that $P^n(A \mid B_1 \equiv b_1, \ldots, B_m \equiv b_m)$ equals

$$\frac{P(A, B_1^n = b_1 n, \ldots, B_m^n = b_m n)}{P(B_1^n = b_1 n, \ldots, B_m^n = b_m n)} \tag{3.19}$$

Due to the fact that $B_1, \ldots, B_m$ form a partition we can apply the law of total probability, compare with Lemma C.6, to segment Eqn. (3.19) yielding

$$\sum_{\substack{1 \leqslant i \leqslant m \\ P(B_i) \neq 0}} \frac{P(A, B_i, B_1^n = b_1 n, \ldots, B_m^n = b_m n)}{P(B_1^n = b_1 n, \ldots, B_m^n = b_m n)} \tag{3.20}$$

It remains to show that each summand in Eqn. (3.20) equals $b_i \cdot P(A|B_i)$. According to Eqn. (3.20) we can henceforth assume that $P(B_i) \neq 0$. By the way the, clause $P(B_i) \neq 0$ in Eqn. (3.20) is only relevant for those cases where $P(B_i) = 0$ and $b_i = 0$ and then allows to omit the respective undefined summand from the sum in Eqn. (3.18) ensuring this sum to be defined also in these cases. In case both $P(B_i)=0$ and $b_i \neq 0$ the bounded F.P. conditionalization $P^n(A|B_1 \equiv b_1, \ldots, B_m \equiv b_m)$ is undefined and the whole Lemma does not apply anyhow. Now, again due to the fact that $B_1, \ldots, B_m$ forms a partition, we can rewrite Eqn. (3.20) as

$$\sum_{\substack{1 \leqslant i \leqslant m \\ P(B_i) \neq 0}} \frac{P(A, B_i, \bigcap_{j \neq i} \overline{B}_j, B_i^n = b_i n, \bigcap_{j \neq i} B_j^n = b_j n)}{P(B_1^n = b_1 n, \ldots, B_m^n = b_m n)} \tag{3.21}$$

We proceed with showing that each summand in Eqn. (3.21) equals $b_i \cdot P(A|B_i)$ for an arbitrary but fixed $1 \leqslant i \leqslant m$ such that $P(B_i) \neq 0$. Due to Corollary 2.9 we can apply Lemma 2.16 to each single event in $B_1, \ldots, B_m$. We need to distinguish the cases $b_i = 0$ and $b_i \neq 0$. In case $b_i = 0$ we know that $P(B_i, B_i^n = b_i n) = 0$ by Eqn. (2.50) so that by Lemma C.5 the whole i-th summand equals zero which equals $0 \cdot P(A|B_i)$ and we are done. In case $b_i \neq 0$ we can apply Eqn. (2.48) one time to shorten and adjust $B_i^n = b_i n$ in presence of B_i and furthermore Eqn. (2.49) $(m-1)$-times to shorten $B_j^n = b_j n$ in presence of $\overline{B}_j$ for all $j \neq i$ which turns the i-th summand into

$$\frac{P(A, B_i, \bigcap_{j \neq i} \overline{B}_j, B_{i(2)} + \cdots + B_{i(n)} = b_i n - 1, \bigcap_{j \neq i} B_{j(2)} + \cdots + B_{j(n)} = b_j n)}{P(B_1^n = b_1 n, \ldots, B_m^n = b_m n)} \tag{3.22}$$

Due to the fact that $B_1, \ldots, B_m$ form a partition we can remove all $\overline{B}_j$ from Eqn. (3.22). Now, due to Lemma C.1 we can cut off $P(AB_i)$ in Eqn. (3.22) yielding

$$\frac{P(AB_i) \cdot P(B_{i(2)} + \cdots + B_{i(n)} = b_i n - 1, \underset{j \neq i}{\cap} B_{j(2)} + \cdots + B_{j(n)} = b_j n)}{P(B_1^n = b_1 n, \ldots, B_m^n = b_m n)} \tag{3.23}$$

Once more, according to Eqn. (3.20) we can assume that $P(B_i) \neq 0$. Therefore, due to $P(AB_i) = P(A|B_i) \cdot P(B_i)$ and, furthermore, Eqn. (2.17) we can turn Eqn. (3.23) into

$$\underbrace{P(A|B_i)}_{\gamma_i} \cdot \underbrace{\frac{\overbrace{P(B_i) \cdot P\left(B_i^{n-1} = b_i n - 1, \underset{j \neq i}{\cap} B_j^{n-1} = b_j n\right)}^{\eta_i}}{P(B_1^n = b_1 n, \ldots, B_m^n = b_m n)}}_{\delta_i} \tag{3.24}$$

Given Eqn. (3.24), it remains to be shown that $\delta_i = b_i$. Now, we can again exploit that $B_1, \ldots, B_2$ form a partition. Due to this we have that $(\langle B_1, \ldots, B_m \rangle_{(i)})_{i \in \mathbb{N}}$ determines multinomial distributions $\mathfrak{M}_{n, P(B_1), \ldots, P(B_m)}$ and $\mathfrak{M}_{n-1, P(B_1), \ldots, P(B_m)}$ in Eqn. (3.24). Due to this togehter with the Lemma's premise that $b_1 + \cdots + b_m = 1$ we can resolve factor δ_i combinatorially, compare with Eqn. (2.26), yielding

$$\frac{P(B_i) \cdot \frac{(n-1)!}{(b_i \cdot n - 1)! \prod_{j \neq i}(b_j n)!} \cdot P(B_i)^{b_i n - 1} \cdot \prod_{j \neq i} P(B_j)^{b_j n}}{\frac{n!}{\prod_{j \in I}(b_j n)!} \cdot \prod_{j \in I} P(B_j)^{b_j n}} \tag{3.25}$$

Finally, after conducting all possible cancellations of $P(B_i)$s, $P(B_j)$s and $(b_j n)!$s in Eqn. (3.25) we arrive at the following:

$$\frac{(n-1)!}{(b_i n - 1)!} \Big/ \frac{n!}{(b_i n)!} = \frac{n! \cdot b_i n}{n \cdot (b_i n)!} \cdot \frac{(b_i n)!}{n!} = b_i \tag{3.26}$$

$\square$

Equation (3.18) is itself called Jeffrey conditionalization, however, it is also called Jeffrey's rule. The side condition in Theorem 3.3 that the frequencies $b_1, \ldots, b_m$ sum up to one is crucial. Without it we could not resolve $P(A \mid B_1 \equiv b_1, \ldots, B_m \equiv b_m)$ combinatorially as a multinomial distribution. Actually, $P(A \mid B_1 \equiv b_1, \ldots, B_m \equiv b_m)$ is undefined for all $b_1, \ldots, b_m$ that violate this condidition. Consider the case that $b_1, \ldots, b_m$ sum up to a value greater than one. Now, we have that the probability of $(\overline{B_1}^n = b_1, \ldots, \overline{B_m}^n = b_m)$ must be zero. The fact that $b_1 + \cdots + b_m > 1$ implies that there must be some repetitions in which more than one of the events B_i occurred, which conflicts with the fact that $B_1, \ldots, B_m$ form a partition. A similar argument applies for the case $b_1 + \cdots + b_m < 1$.

In order to proof Theorem 3.3 we have resolved some F.P. tests combinatorically. This was intuitive and instructive to do so. However, it was not necessary. Alternatively, we can exploit the fact that each condition event of an F.P. conditionalization actually has the *a posteriori* probability that we assign to it, compare with Lemma 2.15, in order to determine these probabilities. Let us turn back to Eqn. (3.24). Remember that it is necessary to show that δ_i equals b_i. Let us consider the numerator η_i in Eqn. (3.24). The idea is simply to apply the proof pattern of shortening/cutting/shifting backwards, compare with Sect. 2.4. We can shift B_i^{n-1} and all B_j^{n-1} in η_i to the right by one resulting into

$$P(B_i) \cdot P(B_{i(2)} + \cdots + B_{i(n)} = b_i n - 1, \underset{j \neq i}{\cap} B_{j(2)} + \cdots + B_{j(n)} = b_j n) \qquad (3.27)$$

Due to Lemma C.1 we have that Eqn. (3.27) equals

$$P(B_i, B_{i(2)} + \cdots + B_{i(n)} = b_i n - 1, \underset{j \neq i}{\cap} B_{j(2)} + \cdots + B_{j(n)} = b_j n) \qquad (3.28)$$

Due to the fact that $B_1, \ldots, B_m$ form a partition we can turn Eqn. (3.28) into

$$P(B_i, \underset{j \neq i}{\cap} \overline{B}_j, B_{i(2)} + \cdots + B_{i(n)} = b_i n - 1, \underset{j \neq i}{\cap} B_{j(2)} + \cdots + B_{j(n)} = b_j n) \qquad (3.29)$$

Next, we can apply Lemma 2.16 to $(B_i, B_{i(2)} + \cdots + B_{i(n)} = b_i n - 1)$ and each of the $(\overline{B}_j, B_{j(2)} + \cdots + B_{j(n)} = b_j n)$ in Eqn. (3.29) yielding

$$P(B_i, \underset{j \neq i}{\cap} \overline{B}_j, B_i^n = b_i n, \underset{j \neq i}{\cap} B_j^n = b_j n) \qquad (3.30)$$

Now, again due to the fact that $B_1, \ldots, B_m$ form a partition we can drop $\cap_{j \neq i} \overline{B}_j$ in Eqn. (3.30) so that it equals $P(B_i, B_1^n = b_1 n, \ldots, B_m^n = b_1 m)$. Given that this equals η_i in Eqn. (3.27), we can can turn δ_i in Eqn. (3.24) into

$$P(B_i, B_1^n = b_1 n, \ldots, B_m^n = b_m n) / P(B_1^n = b_1 n, \ldots, B_m^n = b_m n) \qquad (3.31)$$

Now, we have that Eqn. (3.31) equals $P^n(B_i | B_1 \equiv b_1, \ldots, B_m \equiv b_m)$. Next, we have that this equals b_i; compare with Lemma 2.15. With $\delta_i = b_i$ we have completed our alternative proof of Theorem 3.3.

Chapter 4
Properties of F.P. Conditionalization

In this chapter, we further investigate properties of F.P. conditionalization. We look at probabilities under F.P. semantics, but also at the behavior of conditional probabilities and also conditional expected values under F.P. conditionalization. Table 4.1 provides, at a glance, an overview of properties of F.P. conditionalizations treated in

	Constraint	F.P. Cond.	Probability Value	Reference
(a)	–	$P_{\mathbf{B}}(B_i)$	b_i	Lemma 2.15
(b)	$m=1,\ \mathbf{B}=(B\equiv b)$	$P_{\mathbf{B}}(A)$	$b\cdot P(A|B)+(1-b)\cdot P(A|\overline{B})$	Lemma 3.1
(c)	$B_1,...,B_m$ form a partition	$P_{\mathbf{B}}(A)$	$\sum_{i=1}^m b_i\cdot P(A\,|\,B_i)$	Theorem 3.3
(d)	for arbitrary bound n	$P_{\mathbf{B}}^n(A)$	by recursive computation	Eqn. (4.7) ff.
(e)	for arbitrary bound n	$P_{\mathbf{B}}^n(A)$	by combinatorial computation	Eqn. (4.16)
(f)	–	$P_{\mathbf{B}}(A)$	$\sum P\!\left(A\big|\bigcap_{i\in I}\zeta_i\right)\cdot P\!\left(\bigcap_{i\in I}\zeta_i\big|\bigcap_{i\in I} B_i\equiv b_i\right)$ $(\zeta_i\in\{B_i,\overline{B_i}\})_{i\in I}$ $P\!\left(\bigcap_{i\in I}\zeta_i\right)\neq 0$	Lemma 4.2
(g)	$B_1,...,B_m$ are independent	$P_{\mathbf{B}}(B_{i_1},...,B_{i_k})$	$b_{i_1}b_{i_2}\cdots b_{i_k}$	Lemma 4.3
(h)	$B_1,...,B_m$ are independent	$P_{\mathbf{B}}(B_1,...,B_m)$	$P_{\mathbf{B}}(B_1)\cdots P_{\mathbf{B}}(B_m)$	Eqn. (4.30)
(i)	$B_1,...,B_m$ are independent	$P_{\mathbf{B}}(A)$	$\sum_{I'\subseteq I}\!\left(P\!\left(A\big|\bigcap_{i\in I'}B_i,\bigcap_{i\notin I'}\overline{B_i}\right)\prod_{i\in I'}b_i\prod_{i\notin I'}(1-b_i)\right)$ $P\!\left(\bigcap_{i\in I'}B_i,\bigcap_{i\notin I'}\overline{B_i}\right)\neq 0$	Theorem 4.4
(j)	A is independent of $B_1,...,B_m$	$P_{\mathbf{B}}(A)$	$P(A)$	Lemma 4.5
(k)	$B_1\equiv 100\%,...,B_i\equiv 100\%$ $B_{i+1}\equiv 0\%,...,B_m\equiv 0\%$	$P_{\mathbf{B}}(A)$	$P(A|B_1,...,B_i,\overline{B_{i+1}},...,\overline{B_m})$	Lemma 4.6
(l)	$B_1,...,B_m$ form a partition or $B_1,...,B_m$ are independent $B_1\equiv P(B_1),...,B_m\equiv P(B_m)$	$P_{\mathbf{B}}(A)$	$P(A)$	Lemma 4.7 Lemma 4.8
(m)	$B_1,...,B_m$ form a partition	$P_{\mathbf{B}}(AB_i)$	$b_i\cdot P(A|B_i)$	Lemma 4.9
(n)	$B_1,...,B_m$ form a partition	$P_{\mathbf{B}}(A|B_i)$	$P(A|B_i)$	Theorem 4.10
(o)	$B_1,...,B_m$ are independent	$P_{\mathbf{B}}(A,B_1,...,B_m)$	$b_1\cdots b_m\cdot P(A|B_1,...,B_m)$	Lemma 4.11
(p)	–	$P_{\mathbf{B}}(A|B_1,...,B_m)$	$P(A|B_1,...,B_m)$	Theorem 4.12

Table 4.1 Properties of F.P. conditionalization. Values of various F.P. conditionalizations $P_{\mathbf{B}}(A)=P(A|B_1\equiv b_1,...,B_m\equiv b_m)$, with frequency specifications $\mathbf{B}=B_1\equiv b_1,...,B_m\equiv b_m$ and condition indices $I=\{1,...,m\}$.

© The Author(s) 2017
D. Draheim, *Generalized Jeffrey Conditionalization*, SpringerBriefs in Computer Science, https://doi.org/10.1007/978-3-319-69868-7_4

this book; compare also with Table 4.2 that deals with conditional expected values after F.P. update. Henceforth, we use **B** to denote a whole list of frequency specifications, i.e., $\mathbf{B} = B_1 \equiv b_1, \ldots, B_m \equiv b_m$. Furthermore, we exploit the usual alternative notation of conditional probability $P_B(A)$ also in case of F.P. conditionalization. It is just natural to use $P_\mathbf{B}(\_)$ for F.P. conditionalizations of the form $P(\_|\mathbf{B})$ and leads to a particular succinct notation. In particular, when it comes to the chaining updates as in (n) in Table 4.1, the alternative notation improves readability, i.e., a probability $P_\mathbf{C}(A|B)$ had to be written $P((A|B)|\mathbf{C})$ without alternative notation.

We start the chapter with a discussion of computation of F.P. conditionalizations in Sect. 4.1. After the introduction of a one-step decomposition lemma, we give both a recursive and a combinatorial specification of F.P. computation. In Sect. 4.2, we show how conditional segmentation of F.P. conditionalizations works, see (f) in Table 4.1. Conditional segmentation is the natural generalization of Jeffrey conditionalization as it expresses an F.P. conditionalization as a weighted sum of conditional probabilities. The generalization is in dropping the constraint that the involved condition events form a partition. The strength of conditional segmentation is in highlighting exactly this generalization from Jeffrey to F.P. conditionalization. Otherwise, conditional segmentation is useful as a helping lemma. In that role, it is a rather weak result compared with, e.g., the one-step decomposition lemma provided earlier in this chapter. In Sect. 4.3, we will investigate how F.P. conditionalization behaves with respect to independence. We show that independence of condition events is preserved under F.P. conditionalization, see (g) and (h) in Table 4.1. Next, we combine this preservation of independence with conditional segmentation, see (i) in Table 4.1. As a rather trivial result, we will see that the probability of a target event that is independent of the involved updated events is not affected at all by such an update, see (j) in Table 4.1. In Sect. 4.4, we proof that simultaneous updates of all involved condition events to probabilities of either 100% or 0% amount to classical conditional probabilities, see (k) in Table 4.1. Next, we are able to proof that the target probability of an event is preserved if the condition events are updated with their *a priori* probabilities at least for certain sufficient conditions, i.e., the fact that condition events form a partition and the fact that the condition events are independent, see (l) in Table 4.1.

In Sect. 4.5 we proof that classical conditional probabilities $P(A|B_i)$ are preserved after updates $B_1 \equiv b_1, \ldots, B_m \equiv b_m$ as long as B_i belongs to $B_1, \ldots, B_m$ and $B_1, \ldots, B_m$ form a partition, see (n) in Table 4.1. The fact follow rather immediately from the full Jeffrey conditionalization case in Theorem 3.3. Fact (n) becomes particularly important in Chapt. 5, where we discuss it as a postulate of Jeffrey's probability kinematics. Also, we will see that conditional probabilities under all updated events are preserved by F.P. update, see (p) in Table 4.1.

Finally, in Sect. 4.6 we investigate properties of conditional expected values after F.P. update. Note that the notion of desirability in Ramsey's subjectivism and Jeffrey's logic of decision technically amounts to conditional expected values.

In Sect. 2.2 we have said that we can assume the definedness of F.P. conditionalization in all proofs and argumentations. We have seen that an F.P. conditionalization $P(A|B_1 \equiv b_1, \ldots, B_n \equiv b_n)$ is only defined in case its condition $(B_1 \equiv b_1, \ldots, B_n \equiv b_n)$

has a probability different from zero. Now, we have that the condition's probability becomes zero whenever $P(B_i) = 0$ and $b_i \neq 0$ or $P(B_i) = 1$ and $b_i \neq 1$. Now, the consideration of the extra cases $P(B_i) = 0$ with $b_i = 0$ and $P(B_i) = 1$ with $b_i = 1$ will provide rather little insight in the analysis of F.P. conditionalization, as it shows, e.g., in Sect. 2 in the distinction between the two Lemmas 3.1 and 3.2 or the side condition $P(B_i) \neq 0$ in the sum in Theorem 3.3. At the same time, the maintenance of these cases can amount to substantial extra technical effort. Therefore, it makes sense to restrict our consideration to cases in which $P(B_i) \neq 0$ and $P(B_i) \neq 1$ as expressed by Convention 4. In Convention 5 we remind of a minor detail concerning a usual way of complement set notation. In Convention 6 we remind of usual ways to denote conjunctions of events that we already used before.

Convention 4 (Probability Values for Condition Events)　　　　Given an F.P. conditionalization $P(A|B_1 \equiv b_1, \dots, B_n \equiv b_n)$ we can assume that the condition events are neither almost sure nor almost impossible, i.e., $0 < P(B_i) < 1$ for all B_i in $B_1, \dots, B_m$.

Convention 5 (Complement Set Notation) Given sets S and S' such that we can assume $S' \subseteq S$ as known from the context, we use $s \notin S'$ to denote $s \in S \backslash S'$ as we use $\overline{S}$ to denote $S \backslash S'$.

Convention 6 (Conjunction of Events) Given n events $B_1 \subseteq \Omega$ through $B_n \subseteq \Omega$ we denote the list of these events as $B_1, \dots, B_n$. We denote the *conjunction of the events* $B_1 \cap \dots \cap B_n$ as $B_1 \cdots B_n$ or also as $B_1, \dots, B_n$. If we have $B_1, \dots, B_n$, it is always clear from the context whether we mean a list of events or a conjunction of events. Given a conjunction of events $B_1 \cap \dots \cap B_n$ the notation $B_1, \dots, B_n$ is rather used in probability values as $P(B_1, \dots, B_n)$, whereas the notation $B_1 \cdots B_n$ is rather used in text.

4.1 Computing F.P. Conditionalization

In this section, we see how to compute a bounded F.P. conditionalization of the form $P^n(A|B_1 \equiv b_1, \dots, B_m \equiv b_m)$. Given this, we can compute an F.P. conditionalization with arbitrary exactness via its bounded approximations. We provide a recursive computation and a combinatorial computation. The recursive computation relies on the repeated application of cutting off independent events. The essence of this is grasped in a technical helper Lemma on one-step decomposition. The combinatorial solution generalizes the combinatorial solution of the binomial distribution $\mathfrak{B}_{n,p}(k)$; compare with Def. 2.10. This means that we generalize from the case of sequencing Bernoulli experiments to the case of sequencing multivariate Bernoulli experiments. We do so in generalizing further the combinatorial solution for the case of sequencing bivariate Bernoulli experiments [2, 110, 111, 145, 158]. The recursive vs. the combinatorial solution are not about implementation options. They are simply about two fundamental options of characterizing F.P. conditionalizations.

4.1.1 Recursive Computation of F.P. Conditionalization

With Lemma 4.1 we provide a short, technical helper, which can be turned immediately into a recursive program for the computation of F.P. conditionalization, but
which also proofs useful on other occasions, i.e., in the proofs of Lemma 4.2 and
Theorem 4.4. Basically, the lemma segments an event that involves m frequency
specifications $B_1^n = k_1, \ldots, B_m^n = k_m$ into the 2^m events of possible combinations of
positive and negative occurences of each B_i in $B_1, \ldots, B_m$. Then, it simultaneously
applies Lemma 2.16 to the segmentation, i.e., 2^m times to each summand and then
again m times for each frequency specification $B_i^n = k_i$ in each of the summands.

Lemma 4.1 (One-Step Decomposition) *Given a sequence of i.i.d. multivariate
characteristic random variables* $(\langle A, B_1, \ldots, B_m \rangle_{(i)})_{i \in \mathbb{N}}$, *a number of repetitions*
$n \in \mathbb{N}$, *and (absolute) frequencies* $0 \leqslant k_i \leqslant n$ *for all* $i \in I = \{1, \ldots, m\}$ *we have
the following:*

$$P(A, \bigcap_{i \in I} B_i^n = k_i) = \sum_{\substack{I' \subseteq I \\ \forall i \in I'. k_i \neq 0 \\ \forall i \notin I'. k_i \neq n}} P(A, \bigcap_{i \in I'} B_i, \bigcap_{i \notin I'} \overline{B_i}) \cdot P\left(\bigcap_{i \in I'} B_i^{n-1} = k_i - 1, \bigcap_{i \notin I'} B_i^{n-1} = k_i \right) \quad (4.1)$$

Proof. We start with the left-hand side of Eqn. (4.1), i.e.:

$$P(A, \bigcap_{i \in I} B_i^n = k_i) \tag{4.2}$$

We can segment the event in Eqn. (4.2) with respect to all possible events $\zeta_1, \ldots, \zeta_m$
in which ζ_i can be either B_i or $\overline{B_i}$. In the next equation we will use $I' \subseteq I$ to indicate
exactly those B_i that occur positive in such a combination. We have that Eqn. (4.2)
equals

$$\sum_{I' \subseteq I} P(A, \bigcap_{i \in I'} B_i, \bigcap_{i \notin I'} \overline{B_i}, \bigcap_{i \in I} B_i^n = k_i) \tag{4.3}$$

Due to Lemma 2.9 we can apply Lemma 2.16 m times to each single summand in
Eqn. (4.4) to shorten and adjust $B_i^n = k_i$ for all $i \in I$ resulting into

$$\sum_{\substack{I' \subseteq I \\ \forall i \in I'. k_i \neq 0 \\ \forall i \notin I'. k_i \neq n}} P(A, \bigcap_{i \in I'} B_i, \bigcap_{i \notin I'} \overline{B_i}, \bigcap_{i \in I'} B_{i(2)} + \cdots + B_{i(n)} = k_i - 1, \bigcap_{i \notin I'} B_{i(2)} + \cdots + B_{i(n)} = k_i) \quad (4.4)$$

Due to Lemma C.1 we know that the event A as well as the events B_i and $\overline{B_i}$ for all
$i \in I$ are all independent of all other events involved in Eqn. (4.4), i.e., the events
$B_{i(j)}$ for all $i \in I$ and all $2 \leqslant k \leqslant n$. Therefore and due to Lemma C.4 we have that
the summand in Eqn. (4.4) equals the following for all $I' \subseteq I$:

$$P(A, \bigcap_{i \in I'} B_i, \bigcap_{i \notin I'} \overline{B_i}) \cdot P(\bigcap_{i \in I'} B_{i(2)} + \cdots + B_{i(n)} = k_i - 1, \bigcap_{i \notin I'} B_{i(2)} + \cdots + B_{i(n)} = k_i) \quad (4.5)$$

Finally, due to Eqn. (2.17) we can shift all sums in Eqn. (4.5) to the left by one resulting into

$$P(A, \bigcap_{i \in I'} B_i, \bigcap_{i \notin I'} \overline{B_i}) \cdot P\left(\bigcap_{i \in I'} B_i^{n-1} = k_i - 1, \bigcap_{i \notin I'} B_i^{n-1} = k_i \right) \tag{4.6}$$

$\square$

We will see in due course that it is sufficient to have a program for probabilities $P(B_1^n = k_1, \ldots, B_m^n = k_m)$. Now, with $A = \Omega$ we have that Eqn. (4.1) can be read immediately as a recursive calculation of $P(B_1^n = k_1, \ldots, B_m^n = k_m)$; compare also with the primer on inductive definitions in [44]. We just interpret P on its first and third occurence in Eqn. (4.1) as a program with fixed parameters $B_1, \ldots B_m$ and P (second occurence in Eqn. (4.1)) that receives a number of repetitions n and a list of numbers of occurences $k_1, \ldots, k_m$ as variable parameters. Actually, we are done. Formally, it is not necessary, but might be instructive to write this program out:

$$P(n; (k_i)_{i \in I}) = \sum_{\substack{I' \subseteq I \\ \forall i \in I'. k_i \neq 0 \\ \forall i \notin I'. k_i \neq n}} P(\bigcap_{i \in I'} B_i, \bigcap_{i \notin I'} \overline{B_i}) \cdot P(n - 1; (k_i - 1)_{i \in I'} \cup (k_i)_{i \notin I'}) \tag{4.7}$$

Let us have a closer look at the program in Eqn. (4.7). Obviously, the sum is assumed as a program primitive. Case distinction is provided by the side-conditions of the sum. All the P-expressions are finitely given, as the index set $I = \{1, \ldots, m\}$ is finite. The side condition $0 \leqslant k_i \leqslant n$ for all $i \in I$ in Lemma 4.2 is maintained during the recursive program run. First, k_i cannot become smaller than zero. It can only be decreased in a recursive call if B_i occurs positive in the corresponding summand. However, if B_i occurs positive in a summand and k_i equals zero, the summand is excluded from the sum by the side condition $\forall i \in I'. k_i \neq 0$ so that there is no recursive call to P that might decrease k_i below zero. Second, with a dual argument, it can be seen that $k_i > n$ is not possible for any number of repetitions after program start. For similar reasons we can see that $n = 1$ correctly establishes the termination case. In case $n = 1$ we can analyze that the only possible recursive call to P is $P(0, (0)_{i \in I})$. Now, we can set $P(0, (0)_{i \in I}) = 1$ as it corresponds to $P(B_1^0 = 0, \ldots, B_m^0 = 0)$; compare with Eqn. (2.18).

With $P(n; (k_i)_{i \in I})$ it is straightforward to provide a program for the computation of F.P. conditionalizations. Due to Lemma 2.13 we have that an F.P. conditionalization $P^n(A | B_1 \equiv b_1, \ldots, B_m \equiv b_m)$ equals

$$\frac{P(A, \overline{B_1}^n = b_1, \ldots, \overline{B_m}^n = b_m)}{P(\overline{B_1}^n = b_1, \ldots, \overline{B_m}^n = b_m)} \tag{4.8}$$

Now, we can apply Lemma 4.1 to the numerator of Eqn. (4.8) yielding

$$\sum_{\substack{I' \subseteq I \\ \forall i \in I'.\, b_i \neq 0 \\ \forall i \notin I'.\, b_i \neq 1}} \mathsf{P}(A, \bigcap_{i \in I'} B_i, \bigcap_{i \notin I'} \overline{B_i}) \cdot \frac{\overbrace{\mathsf{P}(\bigcap_{i \in I'} B_i^{n-1} = b_i n - 1, \bigcap_{i \notin I'} B_i^{n-1} = b_i n)}^{\eta}}{\underbrace{\mathsf{P}(B_1^n = b_1 n, \dots, B_m^n = b_m n)}_{\delta}} \tag{4.9}$$

Now, we can use P from Eqn. (4.7) to compute both η and δ in Eqn. (4.9) in order to compute the value of $\mathsf{P}^n(A \mid B_1 \equiv b_1, \dots, B_m \equiv b_m)$.

4.1.2 Combinatorical Computation of F.P. Conditionalization

In this section we want to determine the value of F.P. conditionalization combinatorically. Following the lines of Sect. 4.1.1 we restrict our attention to the probability of events of the form $(B_1^n = k_1, \dots, B_m^n = k_m)$. In case $B_1, \dots, B_m$ form a partition we already know how to determine its probability combinatorially, i.e., as value of the multinomial distribution $\mathfrak{M}_{n, \mathsf{P}(B_1), \dots, \mathsf{P}(B_m)}$; compare with Eqn. (2.26).

Whenever we consider an event $(B_1^n = k_1, \dots, B_m^n = k_m)$ we have an underlying i.i.d. sequence of multivariate characteristic random variables $(\langle B_1, \dots, B_m \rangle_{(i)})_{i \in \mathbb{N}}$. The distribution $\langle k_1, \dots, k_m \rangle \mapsto \mathsf{P}(B_1 = k_1, \dots, B_m = k_m)$ that is determined by a single multivariate characteristic random variable for all $k_i \in \{0, 1\}$ is also called a multivariate Bernoulli distribution in the literature, in accordance with the notion of Bernoulli variable and Bernoulli experiment. Similarly, a bivariate random variable $\langle B, C \rangle$ is said to determine a bivariate Bernoulli experiment.

We start with considering the minimal case of two events, i.e., we consider events of the form $(B^n = k, C^n = m)$. We know that $(B^n = k, C^n = m)$ stands for the event that after n experiment repetitions we have observed exactly k occurrences of event B and, independently, exactly m occurrences of event C. In [2, 145, 158] we can find the combinatorial solution for the probability distribution $(B^n = k, C^n = m)$ for n, k and m that we want to explain in due course; compare also with [110, 111]:

$$\sum_{i=\max\{0, k+m-n\}}^{\min\{k,m\}} \binom{n}{i, k-i, m-i, n-k-m+i} \mathsf{P}(BC)^i \, \mathsf{P}(B\overline{C})^{k-i} \, \mathsf{P}(\overline{B}C)^{m-i} \, \mathsf{P}(\overline{B}\,\overline{C})^{n-k-m+i} \tag{4.10}$$

The binomial coefficient notation in Eqn. (4.10) is a usual one and stands for the binomial coefficient of the multinomial distribution, i.e., we have that

$$\binom{n}{i, k-i, m-i, n-k-m+i} = \frac{n!}{i!\,(k-i)!\,(m-i)!\,(n-k-m+i)!} \tag{4.11}$$

The index range i in Eqn. (4.11) can be narrowed to start from zero, i.e., to the cases $k + m \leqslant n$, because we can always transform the description of an event $(A^n = p)$ into an equivalent description $((\overline{A})^n = n - p)$. Nevertheless, we want to stay with the general case. In order to see the validity of Eqn. (4.10) we need to concentrate on the event BC. We ask, how often this event can occur after n experiment repetitions; compare also with Nilsson probabilistic logic [120, 121]. Let us assume, without loss of generality, that $k \leqslant m$. Then we have that BC can occur at most k times, otherwise the outcome would show more than k occurences of B which would violate $B^n = k$. If $k + m \leqslant n$ we have that BC does not necessarily have to occur and therefore, in this case the number of possible occurrences of BC ranges between 0 and k. In case $k + m > n$ we have that BC must occur at least $k + m - n$ times. This is so, because $k + m - n$ is the length of the minimal overlap of k and m occurrences in n repetitions. Altogether we know that the number i of possible occurrences of BC ranges between $\max\{0, k+m-n\}$ and $\min\{k, m\}$. If we fix an arbitrary number i of occurrences of BC we have that also the number of occurrences of the events $B\overline{C}, \overline{B}C$ and $\overline{B}\,\overline{C}$ are fixed. We have that $B\overline{C}$ must have occurred $k-i$ times, $\overline{B}C$ must have occurred $m-i$ times and $\overline{B}\,\overline{C}$ must have occurred $n-i-(m-i)-(k-i)$ times, which equals $n-m-k+i$. Therefore, we can use i in its possible range to segment the event $(B^n = k, C^n = m)$. Altogether, we have that $(B^n = k, C^n = m)$ equals

$$\sum_{i=\max\{0,k+m-n\}}^{\min\{k,m\}} \mathsf{P}\Big((BC)^n = i, (B\overline{C})^n = k-i, (\overline{B}C)^n = m-i, (\overline{B}\,\overline{C})^n = n-k-m+i\Big) \quad (4.12)$$

We have that the events BC, $B\overline{C}$, $\overline{B}C$ and $\overline{B}\,\overline{C}$ form a partition so that there repetition determines a multinomial distribution; compare with Eqn. (2.26). Therefore, we have that Eqn. (4.12) can be resolved combinatorically as follows:

$$\sum_{i=\max\{0,k+m-n\}}^{\min\{k,m\}} \mathfrak{M}_{n, \mathsf{P}(BC), \mathsf{P}(B\overline{C}), \mathsf{P}(\overline{B}C), \mathsf{P}(\overline{B}\,\overline{C})}(i, k-i, m-i, n-k-m+i) \quad (4.13)$$

Now, Eqn. (4.13) exactly equals Eqn. (4.10). Following the discussion conducted so far, we step from the case of bivariate Bernoulli distributions [2, 145, 158] to the general case of multivariate Bernoulli distributions, based on a list of events $B_1, \ldots, B_m$ of arbitrary length. As an intermediate step, we first rewrite Eqn. (4.10) in a more succinct form with declarative side conditions as follows:

$$\sum_{\substack{\langle i_1, i_2, i_3, i_4 \rangle \in \mathbb{N}_0^4 \\ i_1 + i_2 = k \\ i_1 + i_3 = m \\ i_1 + i_2 + i_3 + i_4 = n}} \binom{n}{i_1, i_2, i_3, i_4} \mathsf{P}(BC)^{i_1}\, \mathsf{P}(B\overline{C})^{i_2}\, \mathsf{P}(\overline{B}C)^{i_3}\, \mathsf{P}(\overline{B}\,\overline{C})^{i_4} \quad (4.14)$$

The correctness of Eqn. (4.14) with respect to $(B^n = k, C^n = m)$ can be seen, even more directly, along the same lines of argumentation as above. Note that the i_j in $i_1, \ldots, i_n$ can take values in $0 \leqslant i_j \leqslant n$ only, because we have the side condition that

$i_1 + i_2 + i_3 + i_4 = n$. This indicates that the sum is finite and Eqn. (4.14) provides a combinatorial solution. As a further intermediate step, we rewrite Eqn. (4.14) with respect to its range index. This is a merely technical issue but makes it easier to see how the generalization works later:

$$\sum_{\substack{\rho:\{BC,B\overline{C},\overline{B}C,\overline{B}\,\overline{C}\} \to \mathbb{N}_0}} \left(\binom{n}{\rho(BC),\rho(B\overline{C}),\rho(\overline{B}C),\rho(\overline{B}\,\overline{C})} \times \prod_{\zeta \in \{BC,B\overline{C},\overline{B}C,\overline{B}\,\overline{C}\}} \mathsf{P}(\zeta)^{\rho(\zeta)} \right) \qquad (4.15)$$

$$\rho(BC) + \rho(B\overline{C}) = k$$
$$\rho(BC) + \rho(\overline{B}C) = m$$
$$\rho(BC) + \rho(B\overline{C}) + \rho(\overline{B}C) + \rho(\overline{B}\,\overline{C}) = n$$

As the next step we generalize Eqn. (4.15) to a combinatorial solution for the testbed probability $\mathsf{P}(B_1^n = k_1, \ldots, B_m^n = k_m)$ with index set $I = \{1, \ldots, m\}$ as follows:

$$\sum_{\substack{\rho: \mathbb{P}(I) \to \mathbb{N}_0}} \left(\frac{n!}{\prod_{I' \subseteq I} \rho(I')!} \times \prod_{I' \subseteq I} \mathsf{P}(\bigcap_{i \in I'} B_i, \bigcap_{i \notin I'} \overline{B}_i)^{\rho(I')} \right) \qquad (4.16)$$

$$\forall i \in I. k_i = \Sigma\{\rho(I') \,|\, I' \subseteq I \wedge B_i \in I'\}$$
$$n = \Sigma\{\rho(I') \,|\, I' \subseteq I\}$$

Let us walk through Eqn. (4.16). Each selection of indices $I' \subseteq I$ is used to determine a combination of positive vs. negative occurrences of B_is. In that sense, a ρ assigns a number of repetitions $\rho(I')$ to each combination of positive and negative occurrences of B_is. The sum in Eqn. (4.16) ranges over all such ρ that fulfill further side conditions. For each ρ and each event B_i in $B_1, \ldots, B_m$ the number of repetitions $\rho(I')$ of all of those combinations I' that show a positive occurrence for the specific event B_i, i.e., $B_i \in I'$, must sum up to the event's specified number of repetitions k_i. Furthermore, for each ρ the number of all repetitions of all combinations I' must sum up to the total number of repetitions n; compare also with Eqn. (4.15).

4.2 Conditional Segmentation of F.P. Conditionalizations

In this section we express each F.P. conditionalization $\mathsf{P}(A|B_1 \equiv b_1, \ldots, B_m \equiv b_m)$ as a weighted sum of conditional expressions, as we did in the case of Jeffrey conditionalization in Chapt. 3. In the case of Jeffrey conditionalization the summands are $\mathsf{P}(A|B_i) \cdot b_i$ for each B_i. Now, in the general case, the probabilities take the following form for each combination $(\zeta_1, \ldots, \zeta_m)$ of positive vs. negative occurrences of B_i:

$$\mathsf{P}(A \,|\, \zeta_1, \ldots, \zeta_m) \cdot \mathsf{P}(\zeta_1, \ldots, \zeta_m \,|\, B_1 \equiv b_1, \ldots, B_m \equiv b_m) \qquad (4.17)$$

Note that the weights in Eqn. (4.17) are again F.P. conditionalizations. As an example, you might want to walk through the segmentation in case of two *a posteriori* specifications $B \equiv b$ and $C \equiv c$:

$$
\begin{aligned}
P(A\,|\,B \equiv b, C \equiv c) = \quad & P(A\,|\,BC) \cdot P(BC\,|\,B \equiv b, C \equiv c) \\
+ & P(A\,|\,B\overline{C}) \cdot P(B\overline{C}\,|\,B \equiv b, C \equiv c) \\
+ & P(A\,|\,\overline{B}C) \cdot P(\overline{B}C\,|\,B \equiv b, C \equiv c) \\
+ & P(A\,|\,\overline{BC}) \cdot P(\overline{BC}\,|\,B \equiv b, C \equiv c)
\end{aligned}
\tag{4.18}
$$

The segmentation Lemma can be useful as a technical lemma in other proofs and argumentations; however, this is actually a minor aspect. The interesting thing is that the Lemma generalizes, in a straightforward manner, the segmentation of Jeffrey conditionalization.

Lemma 4.2 (F.P. Segmentation) *Given an F.P. conditionalization* $P(A|B_1 \equiv b_1, \dots, B_m \equiv b_m)$ *with index set* $I = \{1, \dots m\}$ *we have the following:*

$$
P(A\,|\,B_1 \equiv b_1, \dots, B_m \equiv b_m) = \sum_{\substack{(\zeta_i \in \{B_i, \overline{B}_i\})_{i \in I} \\ P(\bigcap_{i \in I} \zeta_i) \neq 0}} P(A\,|\, \bigcap_{i \in I} \zeta_i) \cdot P(\bigcap_{i \in I} \zeta_i\,|\,B_1 \equiv b_1, \dots, B_m \equiv b_m) \tag{4.19}
$$

Proof. We show the segmentation Lemma for all of its approximations. Henceforth, we us use $I = \{1, \dots, m\}$ to denote the index set of $B_1, \dots, B_m$. Due to Lemma. 2.13 and Lemma 4.1 we have that $P^n(A\,|\,B_1 \equiv b_1, \dots, B_m \equiv b_m)$ equals

$$
\sum_{\substack{I' \subseteq I \\ \forall i \in I'.\, b_i \neq 0 \\ \forall i \notin I'.\, b_i \neq 1}} P(A, \bigcap_{i \in I'} B_i, \bigcap_{i \notin I'} \overline{B}_i) \cdot \frac{P(\bigcap_{i \in I'} B_i^{n-1} = b_i n - 1,\ \bigcap_{i \notin I'} B_i^{n-1} = b_i n)}{P(B_1^n = b_1 n, \dots, B_m^n = b_m n)} \tag{4.20}
$$

Next, we know that the probability of $P(A, \bigcap_{i \in I'} B_i, \bigcap_{i \notin I'} \overline{B}_i)$ in Eqn. (4.20) equals zero in case an event $(\bigcap_{i \in I'} B_i, \bigcap_{i \notin I'} \overline{B}_i)$ has a zero probability. Therefore, we can narrow the sum in Eqn. (4.20) to $P(\bigcap_{i \in I'} B_i, \bigcap_{i \notin I'} \overline{B}_i) \neq 0$. After that we turn the probability $P(A, \bigcap_{i \in I'} B_i, \bigcap_{i \notin I'} \overline{B}_i)$ into the conditional probability $P(A|\bigcap_{i \in I'} B_i, \bigcap_{i \notin I'} \overline{B}_i)$ by transforming each summand in Eqn. (4.20) into

$$
P(A\,|\, \bigcap_{i \in I'} B_i, \bigcap_{i \notin I'} \overline{B}_i) \cdot \frac{\overbrace{P(\bigcap_{i \in I'} B_i, \bigcap_{i \notin I'} \overline{B}_i) \cdot P(\bigcap_{i \in I'} B_i^{n-1} = b_i n - 1, \bigcap_{i \notin I'} B_i^{n-1} = b_i n)}^{\eta}}{P(B_1^n = b_1 n, \dots, B_m^n = b_m n)} \tag{4.21}
$$

Next, let us consider the numerator η in Eqn. (4.21). Due to Eqn. (2.17) we can shift all B_i^{n-1} in η to the right by one resulting into

$$
P(\bigcap_{i \in I'} B_i, \bigcap_{i \notin I'} \overline{B}_i) \cdot P(\bigcap_{i \in I'} B_{i(2)} + \dots + B_{i(n)} = b_i n - 1, \bigcap_{i \notin I'} B_{i(2)} + \dots + B_{i(n)} = b_i n) \tag{4.22}
$$

Due to Lemma C.1 we have that Eqn. (4.22) equals

$$P(\bigcap_{i\in I'} B_i, \bigcap_{i\notin I'} \overline{B_i}, \bigcap_{i\in I'} B_{i(2)} + \cdots + B_{i(n)} = b_i n - 1, \bigcap_{i\notin I'} B_{i(2)} + \cdots + B_{i(n)} = b_i n) \quad (4.23)$$

Next, we can apply Lemma 2.16 to each of the $(B_i, B_{i(2)} + \cdots + B_{i(n)} = b_i n - 1)$ and $(\overline{B_i}, B_{i(2)} + \cdots + B_{i(n)} = b_i n)$ in Eqn. (4.23) resulting into

$$P(\bigcap_{i\in I'} B_i, \bigcap_{i\notin I'} \overline{B_i}, B_1^n = b_1 n, \ldots, B_m^n = b_m n) \quad (4.24)$$

Given that Eqn. (4.24) equals η in Eqn. (4.21), we have that Eqn. (4.20) equals

$$\sum_{\substack{I' \subseteq I \\ \forall i \in I'.b_i \neq 0 \\ \forall i \notin I'.b_i \neq 1 \\ P(\cap_{i\in I'} B_i, \cap_{i\notin I'} \overline{B_i}) \neq 0}} P(A \mid \bigcap_{i\in I'} B_i, \bigcap_{i\notin I'} \overline{B_i}) \cdot \frac{\overbrace{P(\bigcap_{i\in I'} B_i, \bigcap_{i\notin I'} \overline{B_i}, B_1^n = b_1 n, \ldots, B_m^n = b_m n)}^{\eta'}}{P(B_1^n = b_1 n, \ldots, B_m^n = b_m n)} \quad (4.25)$$

Next, we have that $P(\eta')$ in Eqn. (4.25) equals zero for all summands for which there exists some $i \in I'$ with $b_i = 0$ and, simmilarly, for all summands for which there exists an $i \notin I'$ such that $b_i = 1$. Therefore the respective side-conditions can be dropped from the sum in Eqn. (4.25). Therefore, we are done, because we can see that the narrowed sum of Eqn. (4.25) equals the right-hand side of Eqn. (4.19).
□

Conditional segmentation goes beyond the law of total probability; compare with Lemma C.6. Given $P(A \mid B_1 \equiv b_1, \ldots, B_m \equiv b_m)$ with $\mathbf{B} = B_1 \equiv b_1, \ldots, B_m \equiv b_m$ and again $I = \{1, \ldots, m\}$, we have, due to the law of total probability, that $P_\mathbf{B}(A)$ equals

$$\sum_{\substack{I' \subseteq I \\ P(\cap_{i\in I'} B_i, \cap_{i\notin I'} \overline{B_i}) \neq 0}} P_\mathbf{B}(A \mid \cap_{i\in I'} B_i, \cap_{i\notin I'} \overline{B_i}) \cdot P_\mathbf{B}(\cap_{i\in I'} B_i, \cap_{i\notin I'} \overline{B_i}) \quad (4.26)$$

With Eqn. (4.26) we have almost arrived at Eqn. (4.19); however, there is a crucial difference with respect to the first factor that is still an F.P. conditionalization over the updates $\mathbf{B}$.

4.3 Independence and F.P. Conditionalization

F.P. conditionalization preserves independence. Independence transports from the condition variables over to the target event. With the segmentation Lemma 4.2 we have that an F.P. conditionalization can be resolved as a weighted sum of weighted conditional probabilities. With the independence results of this section, we have that the weights are products of *a posteriori probabilities*. Let us consider the case of an F.P. conditionalization $P(A|B \equiv b, C \equiv c)$ with two independent condition variables B and C and let us see what happens in this case:

$$
\begin{aligned}
P(A|B \equiv b, C \equiv c) = \quad & P(A|BC) \cdot bc \\
+ \ & P(A|B\overline{C}) \cdot b(1-c) \\
+ \ & P(A|\overline{B}C) \cdot (1-b)c \\
+ \ & P(A|\overline{B}\,\overline{C}) \cdot (1-b)(1-c)
\end{aligned}
\tag{4.27}
$$

As a matter of course, Jeffrey conditionalization $P(A|B_1 \equiv b_m, B_m \equiv b_m)_J$ can not show such independence result. Jeffrey conditionalization is simply not defined in such cases, as it is only defined for condition variables $B_1, \ldots, B_m$ that form a partition and those are maximally dependent, i.e., the event $B_1 \cdots B_m$ is simply empty. Next, compare Eqn. (4.27) with the conditional segmentation of $P(A|B \equiv b, C \equiv c)$ in Eqn. (4.18), in particular, please consider the weights bc through $(1-b)(1-c)$ in Eqn. (4.27) with the weights $P(BC|B \equiv b, C \equiv c)$ through $P(\overline{B}\,\overline{C}|B \equiv b, C \equiv c)$ in Eqn. (4.18). As we can see by the weights in Eqns. (4.27) and (4.18), the independence result shows particularly nicely in case the target event of an F.P. conditionalization is composed of the condition events:

$$
P(B_1, \ldots, B_m \,|\, B_1 \equiv b_1, \ldots, B_m \equiv b_m) = b_1 b_2 \cdots b_m
\tag{4.28}
$$

Therefore, given independent event $B_1, \ldots, B_m$ we have that they remain independent after update to *a posteriori* probabilities $B_1 \equiv b_1, \ldots, B_m \equiv b_m$, i.e.,

$$
\begin{aligned}
P(B_1, B_2 \ldots, B_m \,|\, B_1 \equiv b_1, \ldots, B_m \equiv b_m) = \quad & P(B_1|B_1 \equiv b_1, \ldots, B_m \equiv b_m) \\
\times \ & P(B_2|B_1 \equiv b_1, \ldots, B_m \equiv b_m) \\
& \vdots \\
\times \ & P(B_m|B_1 \equiv b_1, \ldots, B_m \equiv b_m)
\end{aligned}
\tag{4.29}
$$

To see the result of Eqn. (4.29) even better, let us denote it in a more succinct form:

$$
\boxed{
\begin{aligned}
& P_{\mathbf{B}}(B_1, B_2 \ldots, B_m) = P_{\mathbf{B}}(B_1) \cdot P_{\mathbf{B}}(B_2) \cdots P_{\mathbf{B}}(B_m) \\
& \textit{where } \mathbf{B} = B_1 \equiv b_1, \ldots, B_m \equiv b_m
\end{aligned}
}
\tag{4.30}
$$

4.3.1 Independence of Conditions A Posteriori

Eqn. (4.30) shows very clearly the neutrality of F.P. conditionalization with respect to the independence of condition events. The independence result reinforces our intuition about F.P. conditionalization. The fact that an event actually has the probability that has been assigned to it is one of the first results that we have provided; compare with Lemma 2.15. However, the fact that the probability of an event $P(BC|B \equiv b, C \equiv c)$ equals the product of b and c is not so obvious anymore. Of course, the independence result is the consequence of the rich independence that exists in the underlying i.i.d. sequence of multivariate random variables as expressed in Lemma C.1. Each such i.i.d. sequence is almost everywhere independent. The only independence can show among the random variables of the same repetition. Now, if these random variables are independent themselves, the whole matrix of random variables becomes independent everywhere.

We investigate the important case that the target event of an F.P. conditionalization consists of condition variables in Lemma 4.3. The independence property for this special case is a crucial result. The generalization to arbitrary target events is a straightforward application of conditional segmentation.

Lemma 4.3 (Independence of Conditions A Posteriori)   *Given a sequence of i.i.d. multivariate characteristic random variables* $(\langle B_1, \ldots, B_m \rangle_{(i)})_{i \in \mathbb{N}}$ *such that* $B_1, \ldots, B_m$ *are also mutually independent, as well as a selection* $B_{i_1}, \ldots, B_{i_k}$ *of events from* $B_1, \ldots, B_m$ *we have the following:*

$$P(B_{i_1}, \ldots, B_{i_k} \mid B_1 \equiv b_1, \ldots, B_m \equiv b_m) = \prod_{i \in \{i_1, \ldots, i_k\}} b_i \qquad (4.31)$$

Proof. We show the Lemma for all of its approximations for an arbitrary but fixed bound $n \in \mathbb{N}$. We have that $P^n(B_{i_1}, \ldots, B_{i_k} \mid B_1 \equiv b_1, \ldots, B_m \equiv b_m)$ equals

$$\frac{P(B_{i_1}, \ldots, B_{i_k}, B_1^n = b_1 n, \ldots, B_m^n = b_m n)}{P(B_1^n = b_1 n, \ldots, B_m^n = b_m n)} \qquad (4.32)$$

Let us use $I = \{1, \ldots, m\}$ to denote the index set of $B_1, \ldots, B_m$ and $I' = \{i_1, \ldots, i_k\}$ for the indices of the selection $B_{i_1}, \ldots, B_{i_k}$ of those events. Now, we can apply Lemma 2.16 k times to the numerator in Eqn. (4.32), i.e., to each $(B_i, B_i^n = b_i n)$ for each $i \in I'$ yielding

$$\frac{P(\bigcap_{i \in I'} B_i, \, \bigcap_{i \in I'} B_{i(2)} + \cdots + B_{i(n)} = b_i n - 1, \, \bigcap_{i \notin I'} B_i^n = b_i n)}{P(\bigcap_{i \in I'} B_i^n = b_i n, \, \bigcap_{i \notin I'} B_i^n = b_i n)} \qquad (4.33)$$

Due to Lemma C.4 together with Corollary C.2, Lemma C.1, the premise that $(\langle B_1, \ldots, B_m \rangle_{(i)})_{i \in \mathbb{N}}$ is i.i.d. , the premise that $B_1, \ldots, B_m$ are mutually independent and the trivial fact that B_i can be written as $(B_i^1 = 1)$ for each $i \in I'$ we have that Eqn. (4.33) equals

$$\frac{\left(\prod_{i \in I'} P(B_i) \right) \cdot \left(\prod_{i \in I'} P(B_{i(2)} + \cdots + B_{i(n)} = b_i n - 1) \right) \cdot P(\bigcap_{i \notin I'} B_i^n = b_i n)}{\left(\prod_{i \in I'} P(B_i^n = b_i n) \right) \cdot P(\bigcap_{i \notin I'} B_i^n = b_i n)} \qquad (4.34)$$

After cancelation of $P(\cap_{i \notin I'} B_i^n = b_i n)$ and, again, due to Lemma C.4 we have that Eqn. (4.34) equals

$$\frac{\left(\prod_{i \in I'} P(B_i, B_{i(2)} + \cdots + B_{i(n)} = b_i n - 1) \right)}{\left(\prod_{i \in I'} P(B_i^n = b_i n) \right)} \tag{4.35}$$

Now, we can again apply Lemma 2.16 k times to the numerator in Eqn. (4.35) so that we arrive at

$$\prod_{i \in I'} \frac{P(B_i, B_i^n = b_i n)}{P(B_i^n = b_i n)} \tag{4.36}$$

Now, we have that each factor in Eqn. (4.36) has the form $P(B_i \mid B_i^n = b_i n)$. However, we have that this is the F.P. conditionalization $P^n(B_i \mid B_i \equiv b_i)$. Now, we are actually done, because we know that $P^n(B_i \mid B_i \equiv b_i)$ equals b_i, due to the proof of Lemma 2.15, which proves $P(B_i \mid B_i \equiv b_i)$ for all of its approximations. $\qquad \square$

We say that Lemma 4.3 is about the independence of projective events because they form a selection of the condition variables. Next, we combine this result with conditional segmentation.

4.3.2 Independence and F.P. Segmentation

An F.P. conditionalization $P^n(A \mid B_1 \equiv b_1, \ldots, B_m \equiv b_m)$ with mutually independent condition events $B_1, \ldots, B_m$ can be resolved as the weighted sum of conditional probabilities of the target event A under all possible combinations of positive and negative occurrences of the B_is from $B_1, \ldots, B_m$, weighted by the *a posteriori* probabilities $b_1, \ldots, b_m$, see Theorem 4.4.

Theorem 4.4 (F.P. Segmentation over Independent Conditions) *Given an F.P. conditionalization $P(A \mid B_1 \equiv b_1, \ldots, B_m \equiv b_m)$ such that $B_1, \ldots, B_m$ are mutually independent we have the following:*

$$P(A \mid \bigcap_{i \in I} B_i \equiv b_i) = \sum_{\substack{I' \subseteq I \\ P(\bigcap_{i \in I'} B_i, \bigcap_{i \notin I'} \overline{B}_i) \neq 0}} \left(P(A \mid \bigcap_{i \in I'} B_i, \bigcap_{i \notin I'} \overline{B}_i) \cdot \prod_{i \in I'} b_i \cdot \prod_{i \notin I'} (1 - b_i) \right) \tag{4.37}$$

Proof. Due to Lemma 4.2 we have that $P(A \mid B_1 \equiv b_1, \ldots, B_m \equiv b_m)$ equals

$$\sum_{\substack{I' \subseteq I \\ P(\bigcap_{i \in I'} B_i, \bigcap_{i \notin I'} \overline{B}_i) \neq 0}} \left(P(A \mid \bigcap_{i \in I'} B_i, \bigcap_{i \notin I'} \overline{B}_i) \cdot \underbrace{P(\bigcap_{i \in I'} B_i, \bigcap_{i \notin I'} \overline{B}_i \mid \bigcap_{i \in I} B_i \equiv b_i)}_{\delta} \right) \tag{4.38}$$

Now, let us consider the factor δ in Eqn. (4.38). For each $I' \subseteq I$ we can rewrite δ by turning negative occurrences of $\overline{B}_i$ of B_i into positive occurences while adjusting the corresponding frequencies to $1-b_i$ yielding $\mathsf{P}(\cap_{i \in I}B_i | \cap_{i \in I'}B_i \equiv b_i, \cap_{i \notin I'}B_i \equiv 1-b_i)$. Now, we can apply Lemma 4.3 to this resulting into $\prod_{i \in I'} b_i \cdot \prod_{i \notin I'}(1-b_i)$. Given this, we have that Eqn. (4.38) equals the right-hand side of Eqn. (4.37). $\square$

The special independence Lemma 4.3 is an important one, as it grasps that independence of events is preserved under F.P. conditionalization. In a sense, it is even the crucial one, as Theorem 4.4 is rather an immediate corollary of it. However, it is Theorem 4.4 that establishes, with Eqn. (4.37) a counterpart to Jeffrey's rule. It can be said that Eqn. (4.37) is a Jeffrey-style rule for the case of independence. And this matters. It is fair to say that the cases of independence and partitions are diametral. In case of independent events, an event is not determined (*observationally*) at all by the occurrence of the other events, whereas in case of a partition, an event is completely determined (*observationally*) by the occurrence of the other events.

Jeffrey's rule and the independence rule in Eqn. (4.37) represent two sides of the same coin and, actually, in the extreme case of a single condition event, they fall together. The extreme case of a single event can be considered as the case of two events. This means that it is possible to team an event B together with its complement $\overline{B}$ to turn it into a partition of length two, so that Jeffrey's general rule in Lemma 3.2 becomes applicable. However, we can also team the single event together with the event Ω so that Theorem 4.4 becomes applicable and we have the following:

$$\mathsf{P}(A \,|\, B \equiv b, \overline{B} \equiv 1-b) = \mathsf{P}(A \,|\, B \equiv b, \Omega \equiv 100\%) \qquad (4.39)$$

To see the correctness Eqn. (4.39) let us resolve $\mathsf{P}(A \,|\, B \equiv b, \Omega \equiv 100\%)$ according to Theorem 4.4:

$$\begin{aligned}
\mathsf{P}(A \,|\, B \equiv b, \Omega \equiv 100\%) = \quad & \mathsf{P}(A \,|\, B\Omega) \cdot b \cdot 1 \\
& + \mathsf{P}(A \,|\, B\emptyset) \cdot b \cdot (1-1) \\
& + \mathsf{P}(A \,|\, \overline{B}\Omega) \cdot (1-b) \cdot 1 \\
& + \mathsf{P}(A \,|\, \overline{B}\emptyset) \cdot (1-b) \cdot (1-1)
\end{aligned} \qquad (4.40)$$

With Eqn. (4.40) we have immediately, that $\mathsf{P}(A \,|\, B \equiv b, \Omega \equiv 100\%)$ equals $b \cdot \mathsf{P}(A \,|\, B) + (1-b) \cdot \mathsf{P}(A \,|\, \overline{B})$, i.e., it falls togehter with Jeffrey's basic rule.

4.3.3 Total Independence of Target Events

If the target event A of an F.P. conditionalization $\mathsf{P}(A | B_1 \equiv b_1, \dots, B_n \equiv b_n)$ is mutually independent of the condition events $B_1, \dots, B_m$ the probability of A is not affected by the updates $B_1 \equiv b_1, \dots, B_n \equiv b_n$. This is intuitive and natural. Also in the case of classical conditional probability, we have that $\mathsf{P}(A|B) = \mathsf{P}(A)$ in case A and B are independent. We have that $\mathsf{P}(A|B) = \mathsf{P}(A)$ is often taken as the definition of independence and in a sense it is the definition *per se*, because intuitively it expresses exactly what independence of A from B is about, i.e., that the occurrence of B has no *observable* influence on the probability of A, i.e., no observable influence in the overall experimental setting. Eventually, it is fair to say that $P(A|B_1 \equiv b_1, \ldots, B_m \equiv b_m) = P(A)$ as established by Lemma 4.5 is the natural generalization of $P(A|B) = P(A)$.

Lemma 4.5 (Total Independence of Target Event) *Given an F.P. conditionalization* $P(A|B_1 \equiv b_1, \ldots, B_n \equiv b_n)$ *with index set* $I = \{1, \ldots, m\}$ *such that* $A, B_1, \ldots, B_m$ *are mutually independent we have:*

$$P(A \mid B_1 \equiv b_1, \ldots, B_m \equiv b_m) = P(A) \qquad (4.41)$$

Proof. Due to Lemma 4.2 we have that $P(A|B_1 \equiv b_1, \ldots, B_m \equiv b_m)$ equals

$$\sum_{\substack{(\zeta_i \in \{B_i, \overline{B_i}\})_{i \in I} \\ P(\underset{i \in I}{\cap} \zeta_i) \neq 0}} P(A \mid \underset{i \in I}{\cap} \zeta_i) \cdot P(\underset{i \in I}{\cap} \zeta_i \mid B_1 \equiv b_1, \ldots, B_m \equiv b_m) \qquad (4.42)$$

Due to the fact that $A, B_1, \ldots, B_m$ are mutually independent we have that Eqn. (4.42) equals

$$P(A) \cdot \sum_{\substack{(\zeta_i \in \{B_i, \overline{B_i}\})_{i \in I} \\ P(\underset{i \in I}{\cap} \zeta_i) \neq 0}} P(\underset{i \in I}{\cap} \zeta_i \mid B_1 \equiv b_1, \ldots, B_m \equiv b_m) \qquad (4.43)$$

Due to the fact that the collection of all possible events of the form $\zeta_1 \cdots \zeta_m$ forms a partition and the law of total probability applied to $P(\_ \mid B_1 \equiv b_1, \ldots, B_m \equiv b_m)$ we have that Eqn. (4.43) equals $P(A) \cdot P(\Omega \mid B_1 \equiv b_1, \ldots, B_m \equiv b_m)$ which equals $P(A)$. $\square$

4.4 Updating with Particular Probability Values

Let us come back to the introductory example from Eqn. (2.1). We have that $P(A|B_1 \equiv 100\%, \ldots, B_m \equiv 100\%)$ equals $P(A|B_1, \ldots, B_m)$. Similarly, we have that $P(A|B_1 \equiv 0\%, \ldots, B_m \equiv 0\%)$ equals $P(A|\overline{B}_1, \ldots, \overline{B}_m)$. Lemma 4.6 provides the obvious generalization of these two properties, in that we assign either 0% or 100% to each of the involved conditions.

Lemma 4.6 (Updating with Almost Sure and Almost Impossible Probabilities) *Given an F.P. conditionalization* $P(A|B_1 \equiv b_1, \ldots, B_m \equiv b_m)$ *with the index set* $I = \{1, \ldots, m\}$ *and a subset* $I' \subseteq I$ *of that index set such that* $B_i \equiv 100\%$ *for all* $i \in I'$ *and, furthermore,* $B_i \equiv 0\%$ *for all* $i \notin I'$, *we have the following:*

$$P(A|B_1 \equiv b_1, \ldots, B_m \equiv b_m) = P(A \mid \underset{i \in I'}{\cap} B_i, \underset{i \notin I'}{\cap} \overline{B}_i) \qquad (4.44)$$

Proof. Due to the Lemma's premise on probabilities values, we have that the conditionalization $P(A \mid B_1 \equiv b_1, \dots, B_m \equiv b_m)$ takes the following form:

$$P(A \mid \bigcap_{i \in I'} B_i \equiv 1, \bigcap_{i \notin I'} B_i \equiv 0) \tag{4.45}$$

Henceforth, let n range over all natural numbers such that $nb_i \in \mathbb{N}$ for all $i \in I$. We will use n according to this side condition as range index in limits $lim_{n \to \infty}$ in the sequel. Now, due to the segmentation Lemma 4.2 and, furthermore, the definition of F.P. conditionalization we have that Eqn. (4.45) equals

$$\sum_{\substack{(\zeta_i \in \{B_i, \overline{B_i}\})_{i \in I} \\ P(\bigcap_{i \in I} \zeta_i) \neq 0}} P(A \mid \bigcap_{i \in I} \zeta_i) \cdot \lim_{n \to \infty} \frac{P(\bigcap_{i \in I} \zeta_i, \bigcap_{i \in I'} \overline{B_i^n} = 1, \bigcap_{i \notin I'} \overline{B_i^n} = 0)}{P(\bigcap_{i \in I'} \overline{B_i^n} = 1, \bigcap_{i \notin I'} \overline{B_i^n} = 0)} \tag{4.46}$$

Next, we see that in event of $(\overline{B_i^n} = 1)$ the event $B_{i(j)}$ must have occurred for all repetitions $j \in I$ and therefore it must also have occurred on its first repetition B_i. Similarly, in event of $(\overline{B_i^n} = 0)$ the event $\overline{B_i}$ must have occurred. Therefore, the event $(\bigcap_{i \in I} \zeta_i, \bigcap_{i \in I'} \overline{B_i^n} = 1, \bigcap_{i \notin I'} \overline{B_i^n} = 0)$ in Eqn. (4.46) is empty for all those $(\bigcap_{i \in I} \zeta_i)$ that are different from $(\bigcap_{i \in I'} B_i, \bigcap_{i \notin I'} \overline{B_i})$. Therefore, we can narrow the sum in Eqn. (4.46) to a single summand:

$$P(A \mid \bigcap_{i \in I'} B_i, \bigcap_{i \notin I'} \overline{B_i}) \cdot \lim_{n \to \infty} \underbrace{\frac{P(\bigcap_{i \in I'} B_i, \bigcap_{i \notin I'} \overline{B_i}, \bigcap_{i \in I'} \overline{B_i^n} = 1, \bigcap_{i \notin I'} \overline{B_i^n} = 0)}{P(\bigcap_{i \in I'} \overline{B_i^n} = 1, \bigcap_{i \notin I'} \overline{B_i^n} = 0)}}_{\phi} \tag{4.47}$$

Finally, we can see that the event $(\bigcap_{i \in I'} B_i, \bigcap_{i \notin I'} \overline{B_i}, \bigcap_{i \in I'} \overline{B_i^n} = 1, \bigcap_{i \notin I'} \overline{B_i^n} = 0)$ in Eqn. (4.47) equals $(\bigcap_{i \in I'} \overline{B_i^n} = 1, \bigcap_{i \notin I'} \overline{B_i^n} = 0)$. Therefore, we have that the quotient ϕ in Eqn. (4.47) equals one, which already completes the proof. $\square$

Next, let us come back to updating events with their original probabilities. In Sect. 3.1 we have seen that in case of a single condition event the probability $P(A \mid B \equiv P(B))$ equals $P(A)$. We can at least identify two sufficient conditions that guarantee that this also holds for the update of multiple condition events and their simultaneous update to their *a priori* probabilities. First, in Lemma 4.7 we are able to prove it for the full Jeffrey conditionalization case, i.e., whenever the condition events form a partition. The proof is easy in this case, as it follows rather immediately from Theorem 3.3 and the law of total probability. Next, in Lemma 4.8 we are able to prove it for the case of mutually independent conditions. Again, the proof is straightforward, this time based on Theorem 4.4.

Lemma 4.7 (Updating with Original Probabilities w.r.t Partitions)   *Given an F.P. conditionalization $P(A \mid B_1 \equiv P(B_1), \dots, B_m \equiv P(B_m))$ such that $B_1, \dots, B_m$ form a partition we have the following:*

$$P(A \mid B_1 \equiv P(B_1), \ldots, B_m \equiv P(B_m)) = P(A) \tag{4.48}$$

Proof. Due to Lemma 3.3 and the lemma's premise that $B_1, \ldots, B_m$ form a partition we have that the F.P. conditionalization $P(A \mid B_1 \equiv P(B_1), \ldots, B_m \equiv P(B_m))$ equals

$$\sum_{i=1}^{m} P(B_i) \cdot P(A \mid B_i) \tag{4.49}$$

Now, due to the law of total probability we have that Eqn. (4.49) equals $P(A)$; compare with Lemma C.6. □

Lemma 4.8 (Updating with Original Probabilities w.r.t Independence) *Given an F.P. conditionalization $P(A \mid B_1 \equiv P(B_1), \ldots, B_m \equiv P(B_m))$ such that $B_1, \ldots, B_m$ are mutually independent we have the following:*

$$P(A \mid B_1 \equiv P(B_1), \ldots, B_m \equiv P(B_m)) = P(A) \tag{4.50}$$

Proof. Due to Theorem 4.4 and the lemma's premise that the events $B_1, \ldots, B_m$ are mutually independent we have that $P(A \mid B_1 \equiv P(B_1), \ldots, B_m \equiv P(B_m))$ equals

$$\sum_{\substack{I' \subseteq I \\ P(\bigcap_{i \in I'} B_i, \bigcap_{i \notin I'} \overline{B_i}) \neq 0}} \frac{P(A, \bigcap_{i \in I'} B_i, \bigcap_{i \notin I'} \overline{B_i})}{P(\bigcap_{i \in I'} B_i, \bigcap_{i \notin I'} \overline{B_i})} \cdot \prod_{i \in I'} P(B_i) \cdot \prod_{i \notin I'} P(\overline{B_i}) \tag{4.51}$$

Due to the fact that $B_1, \ldots, B_m$ are independent, we have that $P(\bigcap_{i \in I'} B_i, \bigcap_{i \notin I'} \overline{B_i})$ equals $\prod_{i \in I'} P(B_i) \cdot \prod_{i \notin I'} P(\overline{B_i})$ so that Eqn. (4.51) equals

$$\sum_{\substack{I' \subseteq I \\ P(\bigcap_{i \in I'} B_i, \bigcap_{i \notin I'} \overline{B_i}) \neq 0}} P(A, \bigcap_{i \in I'} B_i, \bigcap_{i \notin I'} \overline{B_i}) \tag{4.52}$$

Next, we have that the collection of events $(\bigcap_{i \in I'} B_i, \bigcap_{i \notin I'} \overline{B_i})$ over all possible sets $I' \subseteq I$ forms a partition. Due to the law of total probability we therefore have that Eqn. (4.52) equals $P(A)$. □

4.5 Conditional Probabilities *A Posteriori*

In this section, we are interested in conditional probabilities after an F.P. conditionalization has been conducted, i.e., after some condition events have been updated with *a posteriori* probabilities. This notion of probability is a particularly important one as it plays a pivotal role in Jeffrey's logic of decision. We will consider a collection of updates $\mathbf{B} = B_1 \equiv b_1, \ldots, B_m \equiv b_m$ and from this update, we gain the probability function $\mathsf{P}(\_|B_1 \equiv b_1, \ldots, B_m \equiv b_m)$ and the probability function $\mathsf{P}^n(\_|B_1 \equiv b_1, \ldots, B_m \equiv b_m)$ for each appropriate number of repetitions n that we denote also as $\mathsf{P_B}$ and $\mathsf{P_B}^n$. Now, given events A and C we will consider the conditional probablity of A given C with respect to $\mathsf{P_B}$ and $\mathsf{P_B}^n$, i.e.,

$$\mathsf{P_B}(A|C) = \frac{\mathsf{P_B}(AC)}{\mathsf{P_B}(C)} \qquad \mathsf{P_B}^n(A|C) = \frac{\mathsf{P_B}^n(AC)}{\mathsf{P_B}^n(C)} \tag{4.53}$$

Note that the probabilities $\mathsf{P_B}(A|C)$ and $\mathsf{P_B}^n(A|C)$ are meant to be applied only to such events A and B for which there is an i.i.d. sequence of multivariate random variables $(\langle A, C, B_1, \ldots, B_m \rangle_{(j)})_{j \in \mathbb{N}}$. Otherwise, the involved probabilities $\mathsf{P_B}(AC)$, $\mathsf{P_B}(C)$, $\mathsf{P_B}^n(AC)$ and $\mathsf{P_B}^n(C)$ would not be well-defined; compare with Defs. 2.12 and 2.14. Fortunately, due to the limit theorems of the reals the considered conditional probabilities $\mathsf{P_B}(A|C)$ and $\mathsf{P_B}^n(A|C)$ are consistent with our definition of F.P. conditionalization in Def. 2.12. We could write $\mathsf{P_B}(A|C)$ equally well as $\mathsf{P}((A|C)|B_1 \equiv b_1, \ldots, B_1 \equiv b_1)$. Here, the point is to see that the inner expression $(A|C)$ stands for a classical conditional probability, whereas the rest of the expression stands for an F.P. conditionalization. Similarly, we could write $\mathsf{P_B}(A|C)$ also as $(\mathsf{P_B})_C(A)$ or even shorter as $\mathsf{P_{B_C}}(A)$.

In the proofs of this book, we only need the unbounded probabilities of the form $\mathsf{P_B}(A|C)$, because we can build our proofs on previous results. However, to grasp the intuition of the new concept of conditional probabilities *a posteriori*, it is helpful to consider the bounded form $\mathsf{P_B}^n(A|C)$. With respect to this, it is helpful to write it out, both in its full and in its compacted form; compare with Def. 2.12 and Lemma 2.14. We have that $\mathsf{P_B}^n(A|C)$ equals

$$\frac{\mathsf{E}(\overline{(AC)}^n | \overline{B_1}^n = b_1, \ldots, \overline{B_m}^n = b_m)}{\mathsf{E}(\overline{C}^n | \overline{B_1}^n = b_1, \ldots, \overline{B_m}^n = b_m)} \qquad \frac{\mathsf{P}(A, C, \overline{B_1}^n = b_1, \ldots, \overline{B_m}^n = b_m)}{\mathsf{P}(C, \overline{B_1}^n = b_1, \ldots, \overline{B_m}^n = b_m)} \tag{4.54}$$

Note that the compacted form in Eqn. (4.54) is already shortened by the testbed probability $\mathsf{P}(\overline{B_1}^n = b_1, \ldots, \overline{B_m}^n = b_m)$, which makes it even more intuitive. But Eqn. (4.54) tells us even more. We see that the introduced notion of conditional probability *a posteriori* is only one among several possible concepts of chaining probabilities. What it does is chaining a classical conditional probability after an F.P. conditionalization. It must not be confused with other possible concepts such as chaining two F.P. conditionalizations after each other or chaining an F.P. conditionalization after a classical conditional probability.

4.5.1 Conservation of Conditional Propabilities for Partitions

As an important result, we have that the *a posteriori* probability of an arbitrary event A given any of the involved condition events B_i equals the *a priori* probability of A given B_i, as long as the condition events $B_1, \ldots, B_m$ form a partition. The result is important as in Jeffrey's probability kinematics approach, it is taken as a postulate to justify Jeffrey conditionalization. We will discuss this in more depth in Chapt. 5. We start with proving a technical helper Lemma 4.9. The lemma is about the property that the *a posteriori* probability $P_B(AB_i)$ equals the *a priori* probability $b_i \cdot P(A|B_i)$, again as long as $B_1, \ldots, B_m$ form a partition. The lemma follows rather immediately from the Jeffrey generalization case in Theorem 3.3. The property that $P_B(AB_i) = b_i \cdot P(AB_i)$ appears as a generalization of the property $P_B(B_i) = b_i$, see (a) in Table 4.1, but it is not. The property $P_B(B_i) = b_i$ is always valid, whereas $P_B(AB_i) = b_i \cdot P(AB_i)$ holds only in case the condition events $B_1, \ldots, B_m$ form a partition. Once Lemma 4.9 is shown, Theorem 4.10 follows immediately. We have turned the property $P_B(AB_i) = b_i \cdot P(A|B_i)$ into an own lemma, because we need it also on other occasions.

Lemma 4.9 (Partitions and Semi-Projective F.P. Conditionalization) *Given an F.P. conditionalization* $P_B(A) = P(A|B_1 \equiv b_1, \ldots, B_m \equiv b_m)$ *such that* $B_1, \ldots, B_m$ *form a partition we have that* $P_B(AB_i) = b_i \cdot P(A|B_i)$ *for all condition events* B_i *in* $B_1, \ldots, B_m$.

Proof. Due to the fact that $B_1, \ldots, B_m$ form a partition and Theorem 3.3 we have that $P_B(AB_i)$ equals

$$\sum_{j=1}^{m} b_j \cdot P(AB_i|B_j) \tag{4.55}$$

We have that $P(AB_i|B_j) = 0$ for all $i \neq j$ so that Eqn. (4.55) equals $b_i \cdot P(AB_i|B_i)$, which equals $b_i \cdot P(A|B_i)$. □

Theorem 4.10 (Preservation of Conditional Probabilities w.r.t. Partitions) *Given an F.P. conditionalization* $P_B(A) = P(A|B_1 \equiv b_1, \ldots, B_m \equiv b_m)$ *such that the events* $B_1, \ldots, B_m$ *form a partition we have that the conditional probability* $P(A|B_i)$ *is preserved after update according to* **B** *for all condition events* B_i *in* $B_1, \ldots, B_m$, *i.e.:*

$$P_B(A|B_i) = P(A|B_i) \tag{4.56}$$

Proof. We have that $P_B(A|B_i)$ equals $P_B(AB_i)/P_B(B_i)$. Due to the lemma's premise that $B_1, \ldots, B_m$ form a partition we can apply Lemma 4.9 to both the the numerator and denumerator of $P_B(AB_i)/P_B(B_i)$, so that it equals $(b_i \cdot P(A|B_i))/(b_i \cdot P(B_i|B_i))$, wich equals $P(A|B_i)$. □

4.5.2 Conservation of Completely Conditional Probabilities

As an other important result, we have that an alternative of the preservation result in Theorem 4.10 holds. We have that the *a priori* probability of an event A conditional on all condition events is preserved after an F.P. update. In case of Theorem 4.10 the considered conditional probability is with respect to a single condition, in case of Theorem 4.12 the conditional probability is with respect to all conditions. First, we start with a technical helper Lemma 4.11 that corresponds to Lemma 4.9. Lemma 4.11 needs the independence of the condition events as a side condition. In that sense it presents the diametral case of Lemma 4.9. The side condition of independence is not needed as a premise for Theorem 4.12 and Lemma 4.11 is not exploited in the proof of Theorem 4.12; however, Lemma 4.11 serves as a helper lemma to proof corresponding properties of conditional expected values *a posteriori* later in Lemma 4.15.

Lemma 4.11 (Independence and Semi-Projective F.P. Conditionalization)
Given an F.P. conditionalization $P_{\mathbf{B}}(A) = P(A|B_1 \equiv b_1,...,B_m \equiv b_m)$ *such that the condition events* $B_1,...,B_m$ *are mutually independent we have the following for all events A:*

$$P_{\mathbf{B}}(A,B_1,...,B_m) = b_1 \cdots b_m \cdot P(A|B_1,...,B_m) \tag{4.57}$$

Proof. Due to the fact that $B_1,\ldots,B_m$ are mutually independent and Theorem 4.4 we have that $P_{\mathbf{B}}(A,B_1,...,B_m)$ equals

$$\sum_{\substack{I' \subseteq I \\ P(\underset{i \in I'}{\cap} B_i, \underset{i \notin I'}{\cap} \overline{B_i}) \neq 0}} \left(P(A,B_1,...,B_m \mid \underset{i \in I'}{\cap} B_i, \underset{i \notin I'}{\cap} \overline{B_i}) \cdot \prod_{i \in I'} b_i \cdot \prod_{i \notin I'} (1-b_i) \right) \tag{4.58}$$

Now, we can see that the event $B_1 \cdots B_m$ determines a single summand in Eqn. (4.58) that can have a value different from zero, so that Eqn. (4.58) equals the probability $P(A,B_1,...,B_m | B_1,...,B_m) \cdot b_1 \cdots b_m$, wich equals $b_1 \cdots b_m \cdot P(A|B_1,...,B_m)$.   □

Theorem 4.12 (Preservation of Completely Conditional Probabilities) *Given an F.P. conditionalization* $P_{\mathbf{B}}(A) = P(A|B_1 \equiv b_1,...,B_m \equiv b_m)$ *we have the following for all events A:*

$$P_{\mathbf{B}}(A|B_1,...,B_m) = P(A|B_1,...,B_m) \tag{4.59}$$

Proof. We have that $P_{\mathbf{B}}(A|B_1,...,B_m)$ equals $P_{\mathbf{B}}(A,B_1,...,B_m)/P_{\mathbf{B}}(B_1,...,B_m)$. Due to Lemma 4.2 we have that this equals

$$\sum_{\substack{(\zeta_i \in \{B_i,\overline{B_i}\})_{i \in I} \\ P(\underset{i \in I}{\cap}\zeta_i) \neq 0}} P(A,B_1,...,B_m | \underset{i \in I}{\cap}\zeta_i) \cdot P_{\mathbf{B}}(\underset{i \in I}{\cap}\zeta_i) \Big/ \sum_{\substack{(\zeta_i \in \{B_i,\overline{B_i}\})_{i \in I} \\ P(\underset{i \in I}{\cap}\zeta_i) \neq 0}} P(B_1,...,B_m | \underset{i \in I}{\cap}\zeta_i) \cdot P_{\mathbf{B}}(\underset{i \in I}{\cap}\zeta_i) \tag{4.60}$$

Now, we have that $B_1 \cdots B_m$ determines a single summand in both of the sums in Eqn. (4.60) that can have values different from zero, so that Eqn. (4.60) equals

$$\frac{P(A,B_1,...,B_m|B_1,...,B_m) \cdot P_{\mathbf{B}}(B_1,...,B_m)}{P(B_1,...,B_m|B_1,...,B_m) \cdot P_{\mathbf{B}}(B_1,...,B_m)} \tag{4.61}$$

After cancelation of $P_{\mathbf{B}}(B_1,...,B_m)$ we have that Eqn. (4.61) equals the conditional probability $P(A,B_1,...,B_m)/P(B_1,...,B_m)$, i.e., $P(A|B_1,...,B_m)$.   □

4.6 F.P. Expectations

In this section we investigate expected values after update from *a priori* to *a posteriori* probabilities. We deal with the general case of conditional expected values of the form $E_{P_B}(v \mid A)$. All the properties of F.P. conditionalizations in Table 4.1 of the form $P_B(A)$ can be generalized to properties of expected values $E_{P_B}(v)$ in a straightforward manner, with v being an arbitrary real-valued discrete random variable. The generalization to properties of conditional expected values of the form $E_{P_B}(v \mid A)$ is not as straightforward any more. When we consider a probability $E_{P_B}(v \mid A)$ we think about the event A as the target event, and the random variable v rather as fixed. This means that, conceptually, the generalization is not about generalizing the target event A of a probability $P_B(A)$ to an arbitrary real-valued random variable, but about moving it into the condition position of an expected value $E_{P_B}(v \mid A)$ in the context of a given random variable v. Technically, this viewpoint makes actually no difference, but conceptually it is very important. In Ramsey's subjectivism and Jeffrey's logic of decision the notion of *desirability* is a crucial concept. Here, the desirability *desA* of an event A is the conditional expected value of an implicitly given utility v under the condition A. That explains the importance of the described viewpoint. We will discuss that further in Sect. 5.3.

Given the described viewpoint, any of the properties listed in Table 4.1 for a probability $P_B(A)$ could be turned into a corresponding property of the expected value $E_{P_B}(v \mid A)$, keeping in mind the respective side-conditions. As an example, we show the result of such generalization for Jeffrey conditionalization, i.e., both the basic and the full case, in rows (B) and (C) in Table 4.2. As you can see by this example, the translation of the properties might end up in rather complex rules that do not add significant conceptual value. However, we are interested in some special properties that express certain preservations and turn out to be, conceptually, particularly interesting; compare with Table 4.2 once more.

	Constraint	F.P. Cond.	Expected Value	Reference
(A)	$B_1, \ldots, B_m$ form a partition	$E_{P_B}(v \mid B_i)$	$E(v \mid B_i)$	Eqn. (4.69)
(B)	$m = 1$, $\mathbf{B} = (B \equiv b)$	$E_{P_B}(v \mid A)$	$\frac{b \cdot P(A\mid B)E(v\mid AB) + (1-b) \cdot P(A\mid \overline{B_1})E(v\mid A\overline{B_1})}{b \cdot P(A\mid B) + (1-b) \cdot P(A\mid \overline{B})}$	–
(C)	$B_1, \ldots, B_m$ form a partition	$E_{P_B}(v \mid A)$	$\frac{\sum_{i=1}^{m} b_i \cdot P(A\mid B_i) \cdot E(v\mid AB_i)}{\sum_{i=1}^{m} b_i \cdot P(A\mid B_i)}$	–
(M)	$B_1, \ldots, B_m$ form a partition	$E_{P_B}(v \mid AB_i)$	$E(v \mid AB_i)$	Lemma 4.13
(N)	$B_1, \ldots, B_m$ form a partition	$E_{P_B(\sqcup B_i)}(v \mid A)$	$E(v \mid AB_i)$	Lemma 4.14
(O)	$B_1, \ldots, B_m$ are independent	$E_{P_B}(v \mid AB_1 \cdots B_m)$	$E(v \mid AB_1 \cdots B_m)$	Lemma 4.15
(P)	$B_1, \ldots, B_m$ are independent	$E_{P_B(\sqcup B_1 \cdots B_m)}(v \mid A)$	$E(v \mid AB_1 \cdots B_m)$	Lemma 4.15

Table 4.2 Properties of F.P. conditional expectations. Values of various F.P.conditionalizations $E_{P_B}(v \mid A)$, with frequency specifications $\mathbf{B} = B_1 \equiv b_1, \ldots, B_m \equiv b_m$ and condition indices $I = \{1, \ldots, m\}$.

Before turning to the cases of F.P. expectation that interest us, let us start with a more standard and less complex scenario, i.e., expected values *a posteriori* in the context of classical conditional probabilities. Given a probability space $(\Omega, \Sigma, \mathsf{P})$ we denote the expected value with respect to P simply as E. This means, we assume P as known from the context. Equally well, we can use the explicit notation E_P for E; compare with Def. B.5. This explains, why we do not need any extra definitions for expected values such as $\mathsf{E}_{\mathsf{P}_\mathbf{B}}(v|A)$ or even $\mathsf{E}_{\mathsf{P}_{\mathbf{B}(\sqcup B_i)}}(v|A)$ that we use in this section. Similarly, the probability $\mathsf{E}_{\mathsf{P}_B}(v|A)$ with respect to a classical conditional probability P_B is well-defined. Intuitively, it expresses the expectation given that an event A occurred, given that also an event B occurred before. Now, we have the following important and intuitively correct rule:

$$\mathsf{E}_{\mathsf{P}_B}(v|A) = \mathsf{E}(AB) \tag{4.62}$$

To see the correctness of Eqn. (4.62) we write it out as

$$\sum_{d \in D} d \cdot \mathsf{P}_B(v = d|A) \tag{4.63}$$

We assume that v in Eqn. (4.63) ranges over a set D, i.e., $v : \Omega \to D$. Now, after resolving all conditional probabilities in Eqn. (4.63) we have that it equals

$$\sum_{d \in D} d \cdot \frac{\mathsf{P}(v = d, AB)/\mathsf{P}(B)}{\mathsf{P}(AB)/\mathsf{P}(B)} \tag{4.64}$$

After cancelation of $\mathsf{P}(B)$ from Eqn. (4.64) we immediately have that it equals $\mathsf{E}_{\mathsf{P}_B}(v|AB)$. Now, let us turn to the more complex scenario of expectations after F.P. conditionalization. If we consider an expectation $\mathsf{P}_\mathbf{B}(v = d, AB_i)$ we assume an i.i.d. sequence $(\langle v : \Omega \to D, A, B_1, \ldots B_m \rangle_{(j)})_{j \in \mathbb{N}}$ of multivariate random variables, so that v is a real-valued and discrete random variable, and A as well as $B_1, \ldots, B_m$ are characteristic random variables. As usual we use $\mathbf{B}$ to denote $B_1 \equiv b_1, \ldots, B_m \equiv b_m$. Now, we have that the conditional expected value $\mathsf{E}(v|AB_i)$ is preserved after update according to $\mathbf{B}$ for all condition events B_i in $B_1, \ldots, B_m$ as long as $B_1, \ldots, B_m$ form a partition. This important property is exactly what is expressed by Lemma 4.13.

Lemma 4.13 (Preservation of Expectations after F.P. Update) *Given an F.P. conditionalization $\mathsf{P}_\mathbf{B}(A) = \mathsf{P}(B_1 \equiv b_1, \ldots, B_m \equiv b_m)$ such that $B_1, \ldots, B_m$ form a partition and a sequence of real-valued discrete random variables $(v_{(j)} : \Omega \longrightarrow D)_{j \in \mathbb{N}}$ such that $(\langle v, A, B_1, \ldots, B_m \rangle_{(j)})_{j \in \mathbb{N}}$ is i.i.d., we have the following:*

$$\mathsf{E}_{\mathsf{P}_\mathbf{B}}(v|AB_i) = \mathsf{E}(v|AB_i) \tag{4.65}$$

Proof. Due to the definition of conditional expected values, we have that $\mathsf{E}_{\mathsf{P}_\mathbf{B}}(v|AB_i)$ equals

$$\sum_{d \in D} d \cdot \frac{\mathsf{P}_\mathbf{B}(v = d, AB_i)}{\mathsf{P}_\mathbf{B}(AB_i)} \tag{4.66}$$

Applying Lemma 4.9 to both the numerator and the denominator of Eqn. (4.66) yields

$$\sum_{d \in D} d \cdot \frac{b_i \cdot P(v=d, A \mid B_i)}{b_i \cdot P(A \mid B_i)} \tag{4.67}$$

Next, it is possible to cancel b_i and $P(B_i)$ from Eqn. (4.67) resulting into

$$\sum_{d \in D} d \cdot \frac{P(v=d, AB_i)}{P(AB_i)} \tag{4.68}$$

Again, due to the definition of conditional expected value, we have that Eqn. (4.68) equals $E(v \mid AB_i)$. □

Lemma 4.13 establishes rule (M) in Table 4.2. See, how it corresponds to rule (m) in Table 4.1. Now, rule (A) in Table 4.2 follows immediately from Lemma 4.13 by setting $A = \Omega$:

$$E_{P_B}(v \mid B_i) = E(v \mid B_i) \tag{4.69}$$

We have that rule (A) in Table 4.2 corresponds to rule (a) in Table 4.1, however, there is a crucial difference. Rule (a) holds unconstrained, whereas rule (A) is only valid in case the conditions $B_1, \ldots, B_m$ form a partition. This is so because the target event has moved into the position of a condition in (A).

As a next step, we generalize rule (n) in Table 4.1 to conditional expectations. The rule (n) was about chaining a classical conditional probability after the occurrence of an F.P. conditionalization. The generalized property considers an F.P. conditionalization with respect to condition events $B_1, \ldots, B_m$, followed by a classical conditional probability projectively over one of the condition events B_i, which is correctly symbolized as $E_{P_B(\_|B_i)}$ for each target event A. Given that the events $B_1, \ldots, B_m$ form a partition, $E_{P_B(\_|B_i)}$ turns out to equal the *a priori* expectation $E(v \mid A)$.

Lemma 4.14 (Preservation of Expectations after Chaining F.P. Updates)
Given an F.P. conditionalization $P_B(A) = P(A \mid B_1 \equiv b_1, \ldots, B_m \equiv b_m)$ such that the events $B_1, \ldots, B_m$ form a partition and, furthermore, a sequence of real-valued discrete random variables $(v_{(j)} : \Omega \longrightarrow D)_{j \in \mathbb{N}}$ such that $(\langle v, A, B_1, \ldots, B_m \rangle_{(j)})_{j \in \mathbb{N}}$ is i.i.d., we have the following for all condition events B_i in $B_1, \ldots, B_m$, i.e.:

$$E_{P_B(\_|B_i)}(v \mid A) = E(v \mid AB_i) \tag{4.70}$$

Proof. Due to Def. B.6 we have that $E_{P_B(\_|B_i)}(v \mid A)$ equals

$$\sum_{d \in D} d \cdot \frac{P_B(v=d, A \mid B_i)}{P_B(A \mid B_i)} \tag{4.71}$$

Next, it is possible to cancel $P(B_i)$ from Eqn. (4.71) resulting into

$$\sum_{d \in D} d \cdot \frac{P_B(v=d, A, B_i)}{P_B(A, B_i)} \tag{4.72}$$

Again, due to Def. B.6 we have that Eqn. (4.72) equals $\mathsf{E}_{\mathsf{P_B}}(v \,|\, AB_i)$ wich equals $\mathsf{E}(v \,|\, AB_i)$ due to Lemma 4.13.   □

Finally, we also generalize rules (o) and (p) from Table 4.1 in Lemma 4.15. The resulting rules (O) and (P) in Table 4.2 are the counterparts of rules (M) and (N) for the case of independence.

Lemma 4.15 (Preservation of Expectations given Independence)   *Given an F.P. conditionalization* $\mathsf{P_B}(A) = \mathsf{P}(A \,|\, B_1 \equiv b_1, ..., B_m \equiv b_m)$ *such that* $B_1, ..., B_m$ *are mutually independent, furthermore, a sequence of real-valued discrete random variables* $(v_{(j)} : \Omega \longrightarrow D)_{j \in \mathbb{N}}$ *such that* $(\langle v, A, B_1, ..., B_m \rangle_{(j)})_{j \in \mathbb{N}}$ *is i.i.d., we have the following:*

$$\mathsf{E}_{\mathsf{P_B}}(v \,|\, A, B_1, ..., B_m) = \mathsf{E}(v \,|\, A, B_1, ..., B_m) \tag{4.73}$$

$$\mathsf{E}_{\mathsf{P_B}(\_|B_1, ..., B_m)}(v \,|\, A) = \mathsf{E}(v \,|\, A, B_1, ..., B_m) \tag{4.74}$$

Proof. (Sketch) Eqns. (4.73) and (4.74) follow from the definition of conditional expectations and Lemma 4.11 together with the premise that $B_1, ..., B_m$ are mutually independent; compare also with the proofs of Lemmas 4.13 and 4.14.   □

Chapter 5
Probability Kinematics and F.P. Semantics

In this chapter, we provide an analysis of some main aspects of Jeffrey's probability kinematics and their relationship to F.P. semantics. We review the postulate of probability kinematics and how it is used to derive Jeffrey's rule of conditionalization. The postulate of probability kinematics turns out to be a consequence in F.P. semantics. Then, we will compare Jeffrey's multi-step chaining of independent events, which is a special case of commutative Jeffrey chaining, with the F.P. segmentation result for independent events. Jeffrey's independent multi-step chaining and independent F.P. segmentation have the same value; however, Jeffrey's multi-step chaining describes a belief function between probability worlds that are, in general, no direct neighbors, whereas the F.P. segmentation result is a property of a single conditionalization. Against this background, we discuss the redesign of the axiomatic basis of probability kinematics. This redesign is oriented towards the result that the counterpart of Jeffrey's postulate for independent events is a consequence in F.P. semantics. Then, we will have a look at desirabilities. Desirabilities are a central notion in Ramsey's subjectivism and Jeffrey's Bayesian framework. We propose to investigate desirabilities after partial update. For example, we will see how the counterpart of Jeffrey's rule looks like for desirabilities.

Furthermore, we will investigate the correspondence between Donkin's principle and Jeffrey's postulate. Donkin's principle is an early postulate on the behavior of a Bayesian-style partial update. It turns out that Donkin's principle and Jeffrey's postulate are equivalent.

In Sect. 5.1 we start with the discussion of the probability kinematics postulate. Sect. 5.2 is about commutative Jeffrey chaining of independent events and its comparison with the respective F.P. segmentation property. Sect. 5.3 leads to a consideration of desirabilities after partial update. Sect. 5.4 deals with the correspondence between Donkin's principle and Jeffrey's postulate.

Remarks on Notation and Terminology

Henceforth, we use both implicit and explicit notation for expressing partial conditionalizations; compare with the notational remarks on p. 4. Given a partial condi-

© The Author(s) 2017

D. Draheim, *Generalized Jeffrey Conditionalization*, SpringerBriefs in Computer Science, https://doi.org/10.1007/978-3-319-69868-7_5

tionalization P_B with frequency specifications $\mathbf{B} = B_1 \equiv b_1, \ldots, B_m \equiv b_m$ we say that the events $B_1 \ldots B_m$ form the update range of the conditionalization. We also call the events $B_1, \ldots, B_m$ simply the updated events although this is a bit sloppy, as not the events, but only their probabilities are updated. On the other hand, given only an update range $\mathbf{C} = C_1, \ldots, C_m$ we use the update notation $P_{!C}$ to denote a partial conditionalization in the respective framework, so that it is not confused with the conditional probability P_C. Furthermore, henceforth, we will call a list of frequency specifications $\mathbf{B} = B_1 \equiv b_1, \ldots, B_m \equiv b_m$ also an update vector.

We continue to refer to the *objects* of probabilities as *events*, uniformly for all frameworks. We do not talk about *propositions* like Keynes, Jaynes, Ramsey and Jeffrey [75, 87, 93, 129], *events/propositions* like Pearl [123, 124] or *properties* like Donkin [38]. With this decision, we do not want to neglect conceptual differences between interpretations of probability on the one hand or the way the laws of probability theory are established by the several frameworks on the other hand. On the contrary, we do not want to get into a maintenance horror with different terminologies when we start to compare different frameworks, exactly in service for better comparability. What we do is a brute-force terminological simplification, which works, because all the frameworks come up with the same laws of probability. Concerning all this, see also the discussion of fundamental approaches to probability theory as well as the fundamental mathematical styles in Appendix A and Appendix B of [78] by Julian Jaynes and, furthermore, the historical account in Chapter VII and Chapter VIII of [93] by John Maynard Keynes. For a convenient explanation by Jeffrey on how he wants the laws of probability to be established, then called probability logic, can be found in [92].

As a minor, technical remark, all the results in this book concerning partitions of the outcome space would also hold, if we would work with a relaxed notion of partition, in which a collection of events $B_1, \ldots, B_m$ is said to form a partition whenever $P(B_i \cap B_j) = 0$ for all $i \neq j$ and $P(B_1 \cup \ldots \cup B_m) = 1$, instead of requiring $B_i \cap B_j = \emptyset$ and $B_1 \cup \ldots \cup B_m = \Omega$. Of course, the relaxed conditions hold whenever the events $B_1, \ldots, B_m$ form a standard partition as sets, but not vice versa. As we have no use for the slightly stronger results based on the relaxed notion in this book, and the maintenance of it would amount to extra technical effort, we stay with the standard notion of partition when we are working with events. The same discussion would apply to the analytical definition of partitions in terms of properties, with $P(B_i \wedge B_j) = 0$ for all $i \neq j$ and $P(B_1 \vee \ldots \vee B_m) = 1$, instead of $B_i \wedge B_j \equiv \bot$ and $B_1 \vee \ldots \vee B_m \equiv \top$.

Jeffrey uses the term *probability kinematics* when he writes about his approach to degree of belief [87]. The word probability kinematics hints to scenarios, in which the probabilities of some events change from some *a priori* probabilities to some new *a posteriori* probabilities. It is natural to use the term probability kinematics for the concrete way that Jeffrey proposes to determine new probabilities in light of such change, i.e., as the weighted sums of conditional probabilities called Jeffrey conditionalization. However, we would like to use it for the whole approach of updating degrees of belief, including its motivation, its postulate, and implications beyond the concrete rule of conditionalization.

5.1 Jeffrey's Postulate and Jeffrey's Rule

5.1.1 Derivation of Jeffrey's Rule

The F.P. semantics of a partial conditionalization $P_{\mathbf{B}}(A)$ concerning a list of updates $\mathbf{B} = B_1 \equiv b_1,...,B_m \equiv b_m$ is given in terms of an i.i.d. multivariate characteristic
random variable $(\langle A, B_1, ... B_m \rangle_{(i)})_{i \in \mathbb{N}}$. Jeffrey conditionalizations instead considers
two worlds, i.e., an *a priori* world of probabilities characterized by a probability
function P and an *a posteriori* world of probabilities characterized by a probability
function $P_{\mathbf{B}}$, again with respect to updates $\mathbf{B} = B_1 \equiv b_1,...,B_m \equiv b_m$. It is assumed
that in both worlds the laws of probability hold. Furthermore, it is essential that
the two probability functions P and $P_{\mathbf{B}}$ are related to each other by a postulate.
The postulate deals with conditionalizations $P(A|B_1 \equiv b_1,...,B_m \equiv b_m)$ in which the
events $B_1,...,B_m$ form a partition. Then, the postulate states that conditional probabilities with respect to one of the updated events are preserved, i.e., we can assume
that $P_{\mathbf{B}}(A|B_i) = P(A|B_i)$ holds for all events A and all events B_i from $B_1,...,B_m$
as longs as $B_1,...,B_m$ form a partition. Persi Diaconis and Sandy Zabell call this
postulate the J-condition [34, 35]. Richard Bradley talks about conservative belief
changes [18, 36]. We call this postulate the probability kinematics postulate, or also
just Jeffrey's postulate for short. We say that Jeffrey's postulate is a bridging statement, as it bridges between the *a priori* world and the *a posteriori* world.

Assuming Jeffrey's postulate, Jeffrey's rule can be derived; compare with [87].
The argumentation goes as follows. We consider an event A and events $B_1,...,B_m$
that form a partition. Due to the fact, that the laws of probability hold in the *a
posteriori* world, we can apply the law of total probability and therefore know that
A's *a posteriori* probability $P_{\mathbf{B}}(A)$ equals

$$P_{\mathbf{B}}(A) = \sum_{\substack{i=1 \\ P(B_i) \neq 0}}^{m} P_{\mathbf{B}}(B_i) \cdot P_{\mathbf{B}}(A|B_i) \qquad (5.1)$$

Next, as the crucial step, we can exploit Jeffrey's postulate that $P_{\mathbf{B}}(A|B_i)$ equals
$P(A|B_i)$ for all events A and condition events B_i. We apply this to each $P_{\mathbf{B}}(A|B_i)$ in
Eqn. (5.1) and this way Eqn. (5.1) is turned into Jeffrey's rule:

$$P_{\mathbf{B}}(A) = \sum_{\substack{i=1 \\ P(B_i) \neq 0}}^{m} P_{\mathbf{B}}(B_i) \cdot P(A|B_i) \qquad (5.2)$$

With Jeffrey's rule, we have a bridge between the *a priori* world and the *a posteriori* world. Jeffrey's rule has been derived from Jeffrey's postulate and the assumption that the laws of probability hold in both worlds. Let us have a closer look at the
postulate. Let us call the postulate's statement that the updated events form a partition the partition requirement. If we want that the derivation of Jeffrey's rule holds,
we must understand the partition requirement as a constraint on both the *a priori*

and the *a posteriori* world. We can take this for granted as part of the postulate or otherwise as consequence of an appropriate background postulate that ensures that logical relations among events are invariant throughout the several considered worlds. And this is exactly what we do. If you do not want to understand the partition requirement as a requirement on both worlds but only on the *a priori* world, you need to change to a relaxed notion of partitions as a first step – compare with the notational remarks on p. 66 – in order to require that all updates values $b_1, \ldots, b_m$ sum up to one as the second step. For better comparability with Donkin's principle later, let us summarize Jeffrey's postulate in Def. 5.1.

Definition 5.1 (Jeffrey's Postulate)   We say that *Jeffrey's postulate* holds **iff** Given an *a priori* probability P, an *a posteriori* probability $P_{!B}$ with a list of events $\mathbf{B} = B_1, \ldots, B_n$, we have that all probabilities conditional on some event from **B** are preserved after update as long as $B_1, \ldots, B_n$ form a partition, i.e., we have that the following holds for all events A:

$$\mathbf{B} \ forms \ a \ partition \implies P_{!\mathbf{B}}(A|B_i) = P(A|B_i) \ for \ all \ B_i \in \mathbf{B} \qquad (5.3)$$

We call Eqn. (5.3) also Jeffrey's preservation property. In a sense Jeffrey's preservation property and Jeffrey's postulate are synonyms. With the term Jeffrey's postulate it is expressed that the preservation property is turned into an axiom for Jeffrey's Bayesian framework. Now, as an important result, we have that the preservation property holds for F.P. conditionalizations; compare with Corollary 5.2.

Corollary 5.2 (F.P.-Jeffrey Entailment) *Jeffrey's preservation property holds for F.P. conditionalizations.*

Proof. Immediate corollary of Theorem 4.10, see also rule (n) in Table 4.1.   □

5.1.2 Jeffrey Conditionalization for Arbitrary Partitions

Jeffrey's postulate considers only F.P. conditionalizations if the updated events form a partition. The derived Jeffrey conditionalization is also only defined for partitions. Jeffrey discusses two directions of generalization. The first one is a brute-force method that is about considering the combinatorial closure of a collection of events. Mathematically, this proposal does not overcome the restriction to partitions. The second approach is about considering chains of conditionalizations with respect to a collection of independent events. This is a crucial consideration, both in its own right and furthermore, because it is an important result with respect to the discussion of chaining Jeffrey conditionalizations. We discuss the latter direction with a comparison to F.P. semantics in Sect. 5.2.

In [87] Jeffrey proposes how to deal with updating events from collections of arbitrary events $B_1, \ldots, B_m$. The idea is to consider all 2^m conjunctions $\zeta_1 \cdots \zeta_m$ of events, where each ζ_i is either the event B_i or its complement $\overline{B_1}$. Jeffrey calls the events $\zeta_1 \cdots \zeta_m$ the atoms of the collection of events $B_1, ..., B_m$. We call the events $\zeta_1 \cdots \zeta_m$ the outcomes of the multivariate Bernoulli trial of $B_1, ..., B_m$. Then, he proposes to make the updates to all events of the form $\zeta_1 \cdots \zeta_m$ instead of to the events B_i from $B_1, ..., B_m$ themselves. Now, the collection of all events of the form $\zeta_1 \cdots \zeta_m$ forms a partition so that Jeffrey's postulate can be exploited for conditionalizations. Instead of computing a value for the conditionalization $P_{\mathbf{B}}(A)$ with the list of frequency specifications $\mathbf{B} = B_1 \equiv b_1, ..., B_m \equiv b_m$ we provide a list of 2^m fresh frequency specifications

$$\mathbf{C} = (\zeta_1 \cdots \zeta_m \equiv b'_{\zeta_1 \cdots \zeta_m})_{\langle \zeta_1 \in \{B_1, \overline{B_1}\}, ..., \zeta_m \in \{B_m, \overline{B_m}\} \rangle}$$

and compute the conditionalization $P_{\mathbf{C}}(A)$ instead of $P_{\mathbf{B}}(A)$ by exploiting Jeffrey's rule as follows:

$$P_{\mathbf{C}}(A) = \sum_{(\zeta_i \in \{B_i, \overline{B_i}\})_{1 \leqslant i \leqslant m}} b'_{\zeta_1 \cdots \zeta_m} \cdot P(A | \zeta_1 \cdots \zeta_m) \tag{5.4}$$

Compare this also with the segmentation Lemma 4.2. This proposal gives a concrete guideline how to deal with arbitrary collections of events, however, in general, always only after the provision of extra frequency specifications. Only with update values for the marginals in $\mathbf{B}$, i.e., without the knowledge of the extra frequency specifications $\mathbf{C}$, the value of $P_{\mathbf{B}}(A)$ is, in general, not defined in the Jeffrey framework. Actually, Jeffrey develops Eqn. (5.4) from a slightly different angle, from scratch; compare with [87] Section 11.5 and Section 11.6. He starts with deriving Jeffrey's rule for the base case of a single frequency specification $B \equiv b, \overline{B} \equiv (1 - b)$ on the basis of Jeffrey's postulate. He then asks for the generalization to lists of frequency specifications of length $n > 2$. As these might be no partitions, he first forms the list of the so-called atoms, which are disjoint, so that he can apply Jeffrey's postulate and arrives at Jeffrey's generalized rule Eqn. (5.4). Conceptually, this makes no difference to the above discussion. What is achieved by Eqn. (5.4) is a generalization to partitions of arbitrary length, but not a generalization to arbitrary collections of events. Of course, in order to derive a rule for partitions of arbitrary length the detour via constructing all outcomes of multivariate Bernoulli trials is not necessary, as Jeffrey's rule can be derived for arbitrary partitions directly from Jeffrey's postulate as we have done in Sect. 5.1.1. This shows that the construction discussed in this section is more about a guideline on how to deal with arbitrary collections of events.

5.2 Commutative Jeffrey Chaining

Chaining of conditionalizations means that we have two update vectors $\mathbf{B}$ and $\mathbf{C}$ and consider a two-step conditionalization, i.e., updates $\mathbf{B}$ first and then updates $\mathbf{C}$, usually denoted as $P_{\mathbf{B}_{\mathbf{C}}}$ or $P_{\mathbf{B}}(\_|\mathbf{C})$, more seldom as $P((\_|\mathbf{C})|\mathbf{B})$. In general, chaining can be done as a series of n successive updates on the basis of n update

vectors $\mathbf{B}_1$ through $\mathbf{B}_n$. We denote chains of conditionalizations also with brack-
eting in an obvious way as $(\cdots(((\mathsf{P}_{\mathbf{B}_1})_{\mathbf{B}_2})_{\mathbf{B}_3})\cdots)_{\mathbf{B}_n}$. When we are dealing with a
Bayesian framework such as probability kinematics, chaining n conditionalizations
is about assuming $n+1$ probability worlds with measures $\mathsf{P}, \mathsf{P}_{\mathbf{B}_1}, (\mathsf{P}_{\mathbf{B}_1})_{\mathbf{B}_2}$ through
$(\cdots(\mathsf{P}_{\mathbf{B}_1})\cdots)_{\mathbf{B}_n}$. It should then also be assumed that the laws of probability hold in
each of the $n+1$ worlds and that bridging statements such as Jeffrey's postulate that
are usually formulated for a single conditionalization hold accordingly for any two
neighbored worlds i and $i+1$ with measures $(\cdots(\mathsf{P}_{\mathbf{B}_1})\cdots)_{\mathbf{B}_i}$ and $(\cdots(\mathsf{P}_{\mathbf{B}_1})\cdots)_{\mathbf{B}_{i+1}}$. The
straightforward approach to define a chained conditionalization $\mathsf{P}_{\mathbf{B}_\mathbf{C}}$ is to define it
as a conditionalization with updates $\mathbf{C}$ in terms of the probability function $\mathsf{P}_\mathbf{B}$ and
strictly on the basis of the respective definition of conditionalization. With classical
conditional probabilities and Jeffrey conditionalization, it is obvious how to do this.
With conditional probabilities, the chained conditional probability $\mathsf{P}_{B_C}(A)$ is just
$\mathsf{P}_B(AC)/\mathsf{P}_B(C)$.

In probability kinematics, chained conditionalization is based on Jeffrey's rule,
as Jeffrey's rule is taken as notion of conditionalization. Given update vectors
$\mathbf{B} = B_1 \equiv b_1,...,B_m \equiv b_m$ and $\mathbf{C} = C_1 \equiv c_1,...,C_{m'} \equiv c_{m'}$ we have that chaining $\mathsf{P}_{\mathbf{B}_\mathbf{C}}$
is defined as follows:

$$\mathsf{P}_{\mathbf{B}_\mathbf{C}}(A) = \sum_{i=1}^{m'} c_i \cdot \mathsf{P}_\mathbf{B}(A|C_i) \tag{5.5}$$

We call the form of chaining in Eqn. (5.5) *Jeffrey chaining*. We can resolve
Eqn. (5.5) further to see that $\mathsf{P}_{\mathbf{B}_\mathbf{C}}(A)$ equals $\sum_{i=1}^{m'} c_i \cdot (\mathsf{P}_\mathbf{B}(AC_i)/\mathsf{P}_\mathbf{B}(C_i))$ so that we
arrive at the following:

$$\mathsf{P}_{\mathbf{B}_\mathbf{C}} = \sum_{i=1}^{m'} c_i \cdot \frac{\sum_{j=1}^{m} b_j \cdot \mathsf{P}(AC_i|B_j)}{\sum_{j=1}^{m} b_j \cdot \mathsf{P}(C_i|B_j)} \tag{5.6}$$

Chaining can be generalized inductively to chains of arbitrary length. We introduce the following alternative notation for Jeffrey chaining for all target events A:

$$\mathsf{P}_{\mathbf{B}_1 \oplus \cdots \oplus \mathbf{B}_n}(A) = (\cdots(\mathsf{P}_{\mathbf{B}_1})\cdots)_{\mathbf{B}_n}(A) \tag{5.7}$$

Jeffrey chaining is non-commutative, i.e., in general we have that $\mathsf{P}_{\mathbf{B} \oplus \mathbf{C}}$ does
not equal $\mathsf{P}_{\mathbf{C} \oplus \mathbf{B}}$. The most basic example for this non-commutativity is chaining
with respect to the same update, albeit with different update values. Here, chaining
Jeffrey generalizations is forgetful, i.e., it is only the last update's impact that is
present in the probability value of the target even. More precisely, given n update
vectors $\mathbf{B}_1 = B_1 \equiv b_{11},...,B_m \equiv b_{1m}$ through $\mathbf{B}_n = B_1 \equiv b_{n1},...,B_m \equiv b_{nm}$ with
same update range $B_1,...,B_m$ but possibly different update values $b_{ij} \neq b_{i'j}$ we have
that $\mathsf{P}_{\mathbf{B}_1 \oplus \cdots \oplus \mathbf{B}_n}(A)$ equals $\mathsf{P}_{\mathbf{B}_n}(A)$. This can be seen immediately by writing out the
case of chaining two conditionalizations $\mathsf{P}_{\mathbf{B}_1 \oplus \mathbf{B}_2}(A)$ for an an arbtirary event A;
compare with Eqn. (5.6):

$$\sum_{i=1}^{m} b_{2_i} \cdot \frac{\sum_{j=1}^{m} b_{1_j} \cdot P(AB_i|B_j)}{\sum_{j=1}^{m} b_{1_j} \cdot P(B_i|B_j)} = \sum_{i=1}^{m} b_{2_i} \cdot \frac{b_{1_i} \cdot P(AB_iB_i)/P(B_i)}{b_{1_i} \cdot P(B_iB_i)/P(B_i)} = P_{\mathbf{B}_2}(A) \qquad (5.8)$$

The forgetfulness of $P_{\mathbf{B}_1 \oplus \cdots \oplus \mathbf{B}_n}(A)$ then follows inductively from Eqn. (5.8). The non-commutativity of Jeffrey conditionalization has been discussed in depth and controversially [37, 39, 57, 61, 101, 135, 149, 154], turning the intention, in the sense of Franz Brentano [21, 22], of conditionalization into the subject of investigation and discourse.

Actually, Jeffrey conditionalization is commutative in important cases. In earlier work [79] Jeffrey has shown commutativity in case of independent update events which is as expressed by the following Theorem 5.3.

Theorem 5.3 ((Jeffrey [79]) Commutative Jeffrey Chaining) *Given that the Jeffrey postulate holds. Given a list of consecutive single frequency specifications $B_1 \equiv b_1$ through $B_n \equiv b_n$ so that the events $B_1, \ldots, B_n$ are mutually independent, we have the following for all index permutations $i_1, \ldots, i_n$ and all target events A:*

$$P_{B_1 \equiv b_1 \oplus \cdots \oplus B_n \equiv b_n}(A) = P_{B_{i_1} \equiv b_{i_1} \oplus \cdots \oplus B_{i_n} \equiv b_{i_n}}(A) \qquad (5.9)$$

Proof. See the proof in [79], Section III.3. The proof is conducted by induction over the length of chaining.   □

The proof of Theorem 5.3 furthermore yields the result that the probability $P_{B_1 \equiv b_1 \oplus \cdots \oplus B_n \equiv b_n}(A)$ equals, for each possible permutation of indices $i_1, \ldots, i_n$, the weighted sum of all outcomes of the multivariate Bernoulli trial of $B_1, \ldots, B_n$, weighted by the product of the corresponding update values as expressed by the following Lemma 5.4.

Lemma 5.4 (Jeffrey Update Belief Chain for Independent Events) *Given that the Jeffrey postulate holds. Given a list of consecutive single frequency specifications $B_1 \equiv b_1, \ldots, B_n \equiv b_n$ with index set $I = \{1, \ldots, n\}$ so that the events $B_1, \ldots, B_n$ are mutually independent, we have the following for all target events A:*

$$P_{B_1 \equiv b_1 \oplus \cdots \oplus B_n \equiv b_n}(A) = \sum_{\substack{I' \subseteq I \\ P(\bigcap_{i \in I'} B_i, \bigcap_{i \notin I'} \overline{B_i}) \neq 0}} \left(P(A \mid \bigcap_{i \in I'} B_i, \bigcap_{i \notin I'} \overline{B_i}) \cdot \prod_{i \in I'} b_i \cdot \prod_{i \notin I'} (1 - b_i) \right) \qquad (5.10)$$

Proof. See Theorem 5.3. See the proof in [79], Section III.3.

The result value of $P_{B_1 \equiv b_1 \oplus \cdots \oplus B_n \equiv b_n}(A)$ in Eqn. (5.10) has exactly the form found for independent F.P. segmentation in Theorem 4.4, see also rule (p) in Table 4.1. Just for the sake of this discussion let us introduce the symbol $\Psi_{\mathbf{B}}(A)$ for the right hand-side of Eqn. (5.10) where **B** stands for the list of updates $B_1 \equiv b_1, \ldots, B_n \equiv b_n$. The difference between the results of Lemma 5.4 and Theorem 4.4 is that the F.P. conditionalization $P(A|B_1 \equiv b_1, \ldots, B_n \equiv b_n)_{\mathrm{F.P.}}$ itself equals $\Psi_{\mathbf{B}}(A)$, i.e., we have that $\Psi_{\mathbf{B}}(A)$ is reached after a single conditionalization, whereas in case of Jeffrey conditionalization it is reached only after n conditionalization steps. In case of Jeffrey conditionalization $P_{B_1 \equiv b_1 \oplus \cdots \oplus B_n \equiv b_n}(A) = \Psi_{\mathbf{B}}(A)$ expresses a belief function between the first world with probability measure P and the $(n+1)$-th world with measure $P_{B_1 \equiv b_1 \oplus \cdots \oplus B_n \equiv b_n}$. Still, with Lemma 5.4 the Jeffrey conditionalization $P(A|B_1 \equiv b_1, \ldots, B_n \equiv b_n)_J$ is undefined, and therefore also does not equal $\Psi_{\mathbf{B}}(A)$. In F.P. semantics we have that $\mathbf{B}$ is an update vector and $B_1, \ldots, B_n$ forms an update range. This is not so with Theorem 5.3 and Lemma 5.4. Here, $\mathbf{B}$ is not an update vector, but a collection of one element update vectors and $B_1, \ldots, B_n$ is not an update range but a collection of one-element update ranges.

In [87] Jeffrey steps from the base case of a single condition event B to arbitrary many condition events $B_1, \ldots, B_m$ by forming the collection of outcomes of multivariate Bernoulli trials of $B_1, \ldots, B_m$ to achieve a partition and make Jeffrey's postulate applicable, under the assumption that values for all the new updates are provided; compare with the explanation in Sect. 5.1.2. In [79] there is an intermediate step between conditionalization for single events and the enforced partitions for arbitrary events, which is about single-step chaining of a list of independent events, i.e., it is exactly about Theorem 5.3. We have that $P_{B_1 \equiv b_1 \oplus \cdots \oplus B_n \equiv b_n}(A) = \Psi_{\mathbf{B}}(A)$ expresses an belief function between the first world and the $(n+1)$-th world. It could be turned into a conditionalization rule $P(A|B_1 \equiv b_1, \ldots, B_n \equiv b_n)_J = \Psi_{\mathbf{B}}(A)$ bridging between the first and the second world. But this is not done. Instead, probability kinematics stays with Jeffrey's rule Eqn. (5.2) as single update rule. This way, conditionalizations with respect to partitions are the only conditionalizations probability kinematics offers. If we want to turn $P_{B_1 \equiv b_1 \oplus \cdots \oplus B_n \equiv b_n}(A) = \Psi_{\mathbf{B}}(A)$ into a further update rule there would be two ways to do so. We could simple add $P(A|B_1 \equiv b_1, \ldots, B_n \equiv b_n)_J = \Psi_{\mathbf{B}}(A)$ as a postulate to probability kinematics. However, with the same right, Jeffrey's rule could be postulated directly without deriving it from Jeffrey's postulate. The second option is to enrich the axiomatic basis by an appropriate further preservation postulate. For F.P. conditionalization we have proven such a property in Theorem 4.12. Due to Theorem 4.12 we have that the following holds in F.P. semantics for each set of partially updated events $\mathbf{B} = B_1, \ldots, B_m$:

$$P_{!\mathbf{B}}(A|B_1, \ldots, B_m) = P(A|B_1, \ldots, B_m) \tag{5.11}$$

Equation (5.11) is a property that is a consequence of F.P. semantics. Now, we could take this as a further postulate for possible world updates teamed together with Jeffrey's postulate in Eqn. (5.3). The postulates Eqn. (5.3) and Eqn. (5.11) would not be in conflict, with Eqn. (5.3) working for partitions and Eqn. (5.11) working for independent conditions. Now, if we had Eqn. (5.11), we could derive the property $P(A|B_1 \equiv b_1, \ldots, B_n \equiv b_n)_J = \Psi_{\mathbf{B}}(A)$ as a valid update rule, completely corresponding to the derivation of Jeffrey's rule seen with Eqns. (5.1) and (5.2). Given a probability $P(A|B_1 \equiv b_1, \ldots, B_n \equiv b_n)_J$ with $\mathbf{B} = B_1 \equiv b_1, \ldots, B_n \equiv b_n$ and $I = \{1, \ldots, n\}$ as usual, we can consider the partition of all outcomes of the multivariate Bernoulli trial of $B_1, \ldots, B_2$ and can apply the law of total probability so that we can turn $P(A|B_1 \equiv b_1, \ldots, B_n \equiv b_n)_J$ into

$$\sum_{\substack{I' \subseteq L \\ P(\bigcap_{i \in I'} B_i, \bigcap_{i \notin I'} \overline{B_i}) \neq 0}} \left(P_\mathbf{B}(A \mid \bigcap_{i \in I'} B_i, \bigcap_{i \notin I'} \overline{B_i}) \cdot P_\mathbf{B}(\bigcap_{i \in I'} B_i, \bigcap_{i \notin I'} \overline{B_i}) \right) \tag{5.12}$$

Now, if we assume Eqn. (5.11) as a postulate we have that each *a posteriori* conditional probability $P_\mathbf{B}(A \mid \bigcap_{i \in I'} B_i, \bigcap_{i \notin I'} \overline{B_i})$ in Eqn. (5.12) can be turned into its corresponding *a priori* conditional probability $P(A \mid \bigcap_{i \in I'} B_i, \bigcap_{i \notin I'} \overline{B_i})$. With respect to the second factors $P_\mathbf{B}(\bigcap_{i \in I'} B_i, \bigcap_{i \notin I'} \overline{B_i})$ we must take a little care. First we need to assume that the statement about independence in Eqn. (5.12) is a statement that holds for both the *a priori* and the *a posteriori* world. We had a similar discussion with respect to the partition requirement in Jeffrey's postulate earlier. Once this is taken for granted we have that each $P_\mathbf{B}(\bigcap_{i \in I'} B_i, \bigcap_{i \notin I'} \overline{B_i})$ can be turned into $\prod_{i \in I'} P_\mathbf{B}(B_i) \times \prod_{i \notin I'} P_\mathbf{B}(\overline{B_i})$ and we are done as we have arrived at $\Psi_\mathbf{B}(A)$ so that $P(A|B_1 \equiv b_1, \ldots, B_n \equiv b_n)_\text{J} = \Psi_\mathbf{B}(A)$. In order to see that we have arrived at $\Psi_\mathbf{B}(A)$ you might even want to write all *a posteriori* values $P_\mathbf{B}(B_i)$ and $P_\mathbf{B}(\overline{B_i})$ as their explicit values b_i resp. $(1 - b_i)$ which is a merely notational issue in the framework of probability kinematics; compare with the notational remarks on p. 4 *ff*.

Last, for the sake of completeness, we summarize once more the correspondence between Jeffrey chaining along a series of independent events and the F.P. segmentation lemma for independent events in Corollary 5.5.

Corollary 5.5 (Jeffrey-F.P. Correspondence for Commutative Chaining) *Given that the Jeffrey postulate holds. Given a list $B_1 \equiv b_1, \ldots, B_n \equiv b_n$ of frequency specifications so that the events $B_1, \ldots, B_n$ are mutually independent, we have that the probability of n-times chaining Jeffrey conditionalizations of single updates from $B_1 \equiv b_1, \ldots, B_n \equiv b_n$ equals the one-time F.P. conditionalization with simultaneous update of all $B_1, \ldots, B_m$, i.e., we have the following for all target events A:*

$$P_{B_1 \equiv b_1 \oplus \cdots \oplus B_n \equiv b_n}(A)_\text{J} = P(A|B_1 \equiv b_1, \ldots, B_n \equiv b_n)_\text{F.P.} \tag{5.13}$$

Proof. Corollary of Lemma 5.4 and Theorem 4.4.

5.3 Jeffrey Desirabilities *A Posteriori*

The notion of desirability is a central concept in the Bayesian framework of Richard Jeffrey. The way Jeffrey establishes and exploits the concept in his framework is rooted in the subjectivism of Frank P. Ramsey. Technically, the desirability of an event A, which is denoted as $des(A)$ by Jeffrey, is the probability value of a real-valued discrete random variable v conditional on the event A, i.e.:

$$des(A) = \mathsf{E}(v|A) \tag{5.14}$$

The random-variable that underlies a desirability is usually called utility. In usual argumentations, the utility is considered as fixed with respect to a concrete decision scenario and therefore can be deemphasized by the implicit notation $des(A)$. The typically discussed decision scenario in Jeffrey's framework is completely determined by a collection of events $A_1, \ldots, A_m$. For the current discussion, it plays no role that the events might be classified with respect to certain concepts, e.g., as acts vs. conditions. Now, the probabilities and desirabilities of all the 2^m so-called atoms $\zeta_1 \cdots \zeta_m$ with ζ_i being either A_i or $\overline{A}_i$ are known; compare also with Sect. 5.1.2. It means, on the other hand, that this information is considered as sufficient. In particular, given an arbitrary event A and the collection of the above 2^m atoms equipped with an arbitrary order as $C_1, \ldots, C_{2^m}$, we have that the following rule holds for the desirability $E(v|A)$ of the event A as an instance of the law of total probability, just because $C_1, \ldots, C_{2^m}$ forms a partition:

$$E(v|A) = \sum_{i=1}^{2^m} E(v|AC_i) \cdot P(C_i|A) \tag{5.15}$$

Now we see that the utility v in Eqn. (5.15) does not contribute to the understanding of the decision scenario and can be dropped without loss resulting, in Jeffrey notation, into

$$des(A) = \sum_{i=1}^{2^m} des(AC_i) \cdot prob(C_i/A) \tag{5.16}$$

In [92] desirabilities are introduced with an explicit utility via Eqn. (5.15); compare also with the original definition quoted in Eqn. (A.1). In [87] they are introduced without a utility, in terms of Eqn. (5.16). To see the correspondence with the steps from the computing procedure of the original definition from [87], that we have quoted in Eqn. (A.2), it is helpful to transform Eqn. (5.16) further into the following form:

$$des(A) = \frac{\sum_{i=1}^{2^m} des(AC_i) \cdot prob(AC_i)}{\sum_{i=1}^{2^m} prob(AC_i)} \tag{5.17}$$

We use both of the notations $des(A)$ and $E(v|A)$ in our discussions, in particular, we speak about expectations given as $E(v|A)$ also as desirabilities.

Now, we are interested in how desirabilities behave under update. In [87] Jeffrey describes the following. After an agent has changed its believe into an event B to a full believe, i.e., has updated its probability to a 100%, we have that the *a posteriori* desirability of another event A conditional on B needs to equal the *a priori* desirability of AB. He symbolizes this as follows:

$$DES(A) = des(AB) \tag{5.18}$$

It is important to understand, that Eqn. (5.18) is a special case, i.e., it is about an update to 100%, which can only be seen by the context of Eqn. (5.18), as Eqn. (5.18) is given in implicit notation. In general, i.e., without that information, Eqn. (5.18) is not valid. The adequate symbolization of Eqn. (5.18) in our explicit notation, including the side condition that B's probability has been changed into a 100%, is the following:

$$\mathsf{E}_{\mathsf{P}_B}(v \,|\, A) = \mathsf{E}(v \,|\, AB) \tag{5.19}$$

With respect to Eqn. (5.19) it is important to point out that P_B is a classical conditional probability $P_B(A) = P(AB)/P(B)$. We have already seen Eqn. (5.19) and discussed its correctness as Eqn. (4.62). This means that so far, we have used $DES(A)$ only to denote the desirability after an update of some event B to 100%. Now, we propose to analyze desirabilities $DES(A)$ after update of some *a priori* probabilities $prob(B_1), \ldots, prob(B_m)$ to some arbitrary *a posteriori* probabilities $PROB(B_1), \ldots, PROB(B_m)$. For this purpose, we can exploit all properties found in Sect. 4.6; compare also with Table 4.2. This way, we can develop the counterpart of Jeffrey's rule for desirabilities. Given an arbitrary event A and a list of updated events $B_1, \ldots, B_m$ we know the desirability $DES(A)$ after update as long as the events $B_1, \ldots, B_m$ form a partition, as follows:

$$DES(A) = \frac{\sum_{i=1}^{m} PROB(B_i) \cdot prob(A|B_i) \cdot des(AB_i)}{\sum_{i=1}^{m} PROB(B_i) \cdot prob(A|B_i)} \tag{5.20}$$

Compare Eqn. (5.20) with Jeffrey's rule in implicit notation in Eqn. (1.6). With $PROB(B_1) = PROB(B) = 100\%$ and $PROB(B_2) = PROB(\overline{B}) = 0\%$ we have that $DES(A)$ in Eqn. (5.20) resolves into $des(AB)$ which is consistent with Eqn. (5.18). Equation (5.20) results from turning rule (C) in Table 4.2 into implicit notation. Eventually, let us restate also the other results of Sect. 4.6 in the implicit Jeffrey notation. Given that the condition events are again $B_1, \ldots B_m$, that the condition events form a partition and that B_i is one of them we have the following:

$$DES(B_i) = des(B_i) \tag{5.21}$$

$$DES(AB_i) = des(AB_i) \tag{5.22}$$

$$DES(A/B_i) = des(AB_i) \tag{5.23}$$

We have that Eqn. (5.21), (5.22) and (5.23) correspond to (A), (M) and (N) in Table 4.2; compare also with Lemma 4.13 and Lemma 4.14. In particular, please compare Eqns. (5.22) and (5.23) with the kinematics postulate in Def. 5.1. The usage of the notation $DES(A/B_i)$ for $\mathsf{E}_{\mathsf{P}_{B \sqcup B_i}}(v|A)$ in Eqn. (5.23) is not immediately given, i.e., it is our choice to use this notation. It is Eqn. (5.23) that can be considered the counterpart of Jeffrey's postulate for desirabilities.

5.4 The Jeffrey-Donkin Correspondence

Donkin's principle is a postulate on the behavior of a Bayesian-style partial update formulated by William F. Donkin [38]; compare also with remarks of George Boole in [10] and John Maynard Keynes in [93]. It turns out that Donkin's principle is equivalent to Jeffrey's probability kinematics postulate. Basically, it is this correspondence what this section is about. Let us start with the original statement of Donkin's principle in [38]:

> "[..] I think that considerable advantage might be gained by the introduction of the following preliminary theorem, which, if it ought not rather to be called an axiom, is certainly as evident before as after any proof which can be given of it.
> *Theorem.* – If there be any number of mutually exclusive hypotheses, $h_1, h_2, h_3 \ldots$ of which the probabilities relative to a particular state of information are $p_1, p_2, p_3 \ldots$ and if new information be gained which changes the probabilities of some of them, suppose of h_{m+1} and all that follow, *without having otherwise any reference to the rest*, then the probabilities of these latter have the *same ratios* to one another, *after* the new information, that they had *before*; that is,   (5.24)
>
> $$p_1' : p_2' : p_3' : \cdots : p_m' = p_1 : p_2 : p_3 : \cdots : p_m,$$
>
> where the accented letters denote the values after the new information has been acquired."[38]

We restate Donkin's principle in Def. 5.6 by introducing probability functions, plus the slight difference that we make explicit the list of partially updated events $h_{m+1}, \ldots$ as a list of events $B_1, \ldots, B_n$ and the list of investigated events $h_1, \ldots, h_m$ as a list of events $C_1, \ldots, C_m$, which is just a matter of convenience in order to distinguish them more quickly in argumentations.

Definition 5.6 (Donkin's Principle) We say that *Donkin's principle* holds **iff** Given an *a priori* probability P, an *a posteriori* probability $\mathsf{P}_{!\mathbf{B}}$ with a list of events $\mathbf{B} = B_1, \ldots, B_n$ and a list of *(affected)* events $\mathbf{C} = C_1, \ldots, C_m$ so that all events $C_1, \ldots, C_m, B_1, \ldots, B_n$ are mutually exclusive, we have that all probability ratios of events from $\mathbf{C}$ are preserved after update, i.e., we have that the following holds for all indices $1 \leqslant p \leqslant m$ and $1 \leqslant q \leqslant m$:

$$\mathsf{P}_{!\mathbf{B}}(C_p)/\mathsf{P}_{!\mathbf{B}}(C_q) = \mathsf{P}(C_p)/\mathsf{P}(C_q) \tag{5.25}$$

Next, we reformulate Donkin's principle in terms of updating conditions that form a partition. This is a merely presentational issue, in order to make the comparison with Jeffrey conditionalization more convenient. No generality is lost, as a collection of mutually exclusive conditions $B_1, \ldots, B_m$ can always be turned into a partition by adding the event $\overline{B_1 \cdots B_m}$ with update value $(1 - (\mathsf{P}_{!\mathbf{B}}(B_1) + \cdots + \mathsf{P}_{!\mathbf{B}}(B_m)))$ to it. Actually, the original formulation of Donkin's principle is more convenient and easier to grasp as it stands, however, it is harder to deal with in proofs. Later, we will see that Donkin's principle can be stated in a more parsimonious, equivalent way, as the condition that $C_1, \ldots, C_2$ are mutually exclusive can be relaxed to the condition that each of them entails $\overline{B_1 \cdots B_m}$. That this is a relaxation can also be seen more directly with the reformulated version. Next, we will prove that Donkin's principle holds whenever we assume Jeffrey's probability kinematics postulate; compare with Def. 5.1.

Definition 5.7 (Donkin's Principle in Terms of Partitions) We say that *Donkin's principle* holds **iff** Given an *a priori* probability P, an *a posteriori* probability $P_{!B}$ with a list of events $\mathbf{B} = B_1, \ldots, B_n$ that form a partition and a list of events $\mathbf{C} = C_1, \ldots, C_m$ that are mutually exclusive so that there is an event B in $\mathbf{B}$ that is exhaustive w.r.t to the events $\mathbf{C}$, i.e., $C_1 \cup \cdots \cup C_n \subseteq B$, we have that the following holds for all indices $1 \leqslant p \leqslant m$ and $1 \leqslant q \leqslant m$:

$$P_{!B}(C_p)/P_{!B}(C_q) = P(C_p)/P(C_q) \tag{5.26}$$

Lemma 5.8 (Jeffrey-Donkin Entailment) *If we assume that Jeffrey's postulate holds we can also assume Donkin's principle.*

Proof. Given a list of events $\mathbf{B} = B_1, \ldots, B_n$ that form a partition, a list of events $\mathbf{C} = C_1, \ldots, C_m$ that are mutually exclusive so that there is an event B in $\mathbf{B}$ that is exhaustive w.r.t to the events $\mathbf{C}$, as well as indices $1 \leqslant p \leqslant m$ and $1 \leqslant q \leqslant m$. We need to show:

$$P_{!B}(C_p)/P_{!B}(C_q) = P(C_p)/P(C_q) \tag{5.27}$$

Due to the fact that B is exhaustive w.r.t to all events in $\mathbf{C}$, we have that both $C_p B = C_p$ and $C_q B = C_q$. Therefore we immediately have that $P_{!B}(C_p)/P_{!B}(C_q)$ equals $P_{!B}(C_p B)/P_{!B}(C_q B)$, which equals

$$P_{!B}(C_p|B)/P_{!B}(C_q|B) \tag{5.28}$$

Due to the premise we can assume Jeffrey's postulate so that $P_{!B}(C_p|B) = P(C_p|B)$ as well as $P_{!B}(C_q|B) = P(C_q|B)$. Therefore, Eqn. (5.28) equals $P(C_p|B)/P(C_q|B)$, which equals $P(C_p B)/P(C_q B)$, which equals $P(C_p)/P(C_q)$, again due to the fact that B is exhaustive w.r.t the events in $\mathbf{C}$. $\square$

As a crucial step towards proving the correspondence between Jeffrey's postulate and Donkin's principle we proof a technical helper lemma that gives the value $P_{!B}(B_i) \cdot P(A|B_i)$ to the conditionalization $P_{!B}(AB_i)$. We have proven the same result for F.P. conditionalization in Lemma 4.9; compare with rule (m) in Table 4.1, where we have called the conditionalization of the form $P_{!B}(AB_i)$ semi-projective, as it is projective in terms of the condition B_i.

Lemma 5.9 (Semi-Projective Donkin's Conditionalization) *Given that we can assume Donkin's principle. Then, given a list of events $\mathbf{B} = B_1, \ldots, B_n$ that form a partition, we have the following for all events A and all conditions from B_i from $\mathbf{B}$:*

$$P_{!B}(AB_i) = P_{!B}(B_i) \cdot P(A|B_i) \tag{5.29}$$

Proof. Given an arbitrary event A and an arbitrary condition event B_i from $B_1, \ldots, B_m$. Now, we have that the events AB_i and $\overline{A}B_i$ form a partition of B_i. This means, immediately, that AB_i and $\overline{A}B_i$ are mutually exclusive and also, that B_i is exhaustive w.r.t. $AB_i \cup \overline{A}B_i = B_i$. Therefore, we can apply Donkin's principle to the scenario so that the following relationship holds true:

$$P_{!B}(AB_i)/P_{!B}(\overline{A}B_i) = P(AB_i)/P(\overline{A}B_i) \tag{5.30}$$

Again due to the fact that the events AB_i and $\overline{A}B_i$ form a partition of B_i, we have that $P_{!B}(\overline{A}B_i)$ equals $P_{!B}(B_i) - P_{!B}(AB_i)$. Therefore, we can turn Eqn. (5.30) into

$$\frac{P_{!B}(AB_i)}{P_{!B}(B_i) - P_{!B}(AB_i)} = \frac{P(AB_i)}{P(\overline{A}B_i)} \tag{5.31}$$

Next, we conduct a series of equivalent transformation with respect to Eqn. (5.31) as follows:

$$P_{!B}(AB_i) = \frac{P(AB_i)}{P(\overline{A}B_i)}(P_{!B}(B_i) - P_{!B}(AB_i))$$

$$\Longleftrightarrow \quad P_{!B}(AB_i) = \frac{P(AB_i)}{P(\overline{A}B_i)}P_{!B}(B_i) - \frac{P(AB_i)}{P(\overline{A}B_i)}P_{!B}(AB_i)$$

$$\Longleftrightarrow \quad P_{!B}(AB_i) + P_{!B}(AB_i)\frac{P(AB_i)}{P(\overline{A}B_i)} = \frac{P(AB_i)}{P(\overline{A}B_i)}P_{!B}(B_i)$$

$$\Longleftrightarrow \quad P_{!B}(AB_i)\left(1 + \frac{P(AB_i)}{P(\overline{A}B_i)}\right) = \frac{P(AB_i)}{P(\overline{A}B_i)}P_{!B}(B_i)$$

$$\Longleftrightarrow \quad P_{!B}(AB_i)\left(\frac{P(\overline{A}B_i) + P(AB_i)}{P(\overline{A}B_i)}\right) = \frac{P(AB_i)}{P(\overline{A}B_i)}P_{!B}(B_i) \tag{5.32}$$

Again due to the fact that the events AB_i and $\overline{A}B_i$ form a partition of B_i, we have that $P(\overline{A}B_i) + P(AB_i)$ equals $P(B_i)$. Therefore, we can turn Eqn. (5.32) into

$$P_{!B}(AB_i)\frac{P(B_i)}{P(\overline{A}B_i)} = \frac{P(AB_i)}{P(\overline{A}B_i)}P_{!B}(B_i) \tag{5.33}$$

Next, we multiply both sides in Eqn. (5.33) by $P(\overline{A}B_i)/P(B_i)$ so that $P_{!B}(AB_i)$ turns out to equal $(P(AB_i)/P(B_i)) \cdot P_{!B}(B_i)$, which equals $P_{!B}(B_i) \cdot P(A|B_i)$. $\quad\square$

With Lemmas 5.8 and 5.9 we have everything to prove the equivalence of Jeffrey's postulate and Donkin's principle in Theorem 5.10.

Theorem 5.10 (Jeffrey-Donkin Correspondence)   *Jeffrey's postulate and Donkin's principle are equivalent.*

Proof. The fact that the probability kinematics postulate entails Donkin's principle has been shown in Lemma 5.8. With respect to the opposite direction we assume a list of events $\mathbf{B} = B_1, \ldots, B_n$ that form a partition as given. We consider arbitrary but fixed events A and B_i. First, we have that $P_{!B}(A|B_i)$ equals $P_{!B}(AB_i)/P_{!B}(B_i)$. Now, we can apply Lemma 5.9 yielding that $P_{!B}(AB_i)/P_{!B}(B_i)$ equals $(P_{!B}(B_i) \cdot P(A|B_i)))/P_{!B}(B_i)$, which equals $P(A|B_i)$ as desired. $\quad\square$

As we will proof now, the conditions in Donkin's principle can be relaxed with respect to the affected events, i.e., the requirement that the affected events need to be mutually exclusive can be dropped, see Corollary 5.11.

Corollary 5.11 (Donkin's Principle with Relaxed Conditions) *Given that we can assume Donkin's principle. Then, given a probability* P, *a probability* $P_{!B}$ *with a list of events* $\mathbf{B} = B_1, \ldots, B_n$ *that form a partition and a list of events* $\mathbf{C} = C_1, \ldots, C_m$ *so that there is an event B in* $\mathbf{B}$ *that is exhaustive w.r.t to the events* $\mathbf{C}$, *we have the* $P_{!B}(C_p)/P_{!B}(C_q) = P(C_p)/P(C_q)$ *for all indices* $1 \leqslant p \leqslant m$ *and* $1 \leqslant q \leqslant m$.

Proof. Corollary of Lemma 5.9. Given arbitrary but fixed indices $1 \leqslant p \leqslant m$ and $1 \leqslant q \leqslant m$. Due to the lemma's premise that B is exhaustive w.r.t. $\mathbf{C}$ we have that $P_{!B}(C_p)/P_{!B}(C_q)$ equals $P_{!B}(C_pB)/P_{!B}(C_qB)$. Now, due to Lemma 5.9, this equals $(P_{!B}(B) \cdot P(C_pB))/(P_{!B}(B) \cdot P(C_qB))$, which equals $P(C_pB)/P(C_qB)$, which equals $P(C_p)/P(C_q)$, again due to the premise that $\mathbf{B}$ is exhaustive w.r.t. $\mathbf{C}$. $\qquad\square$

With Corollary 5.11 we have stated Donkin's principle in terms of more relaxed conditions. However, the principle stays the same; both versions are equivalent. The opposite direction of Corollary 5.11 is even trivial, i.e., once Donkin's principle is postulated in its shorter form it also applies to the more specific scenarios, in which the affected events are mutually exclusive. In the proof of the Jeffrey-Donkin entailment in Lemma 5.10 we can already see that the exclusiveness of affected events can be dropped, i.e., it is simply not exploited in the proof. Of course, Donkin's principle also holds for F.P. conditionalization. This is immediately so due to Theorem 4.10, which proofs that Jeffrey's postulate is a consequence of F.P. semantics, see also rule (n) in Table 4.1, together with the Jeffrey-Donkin entailment that we have just discussed.

Appendix A
Bibliographic Notes

Richard C. Jeffrey's Writings

The derivation of Jeffrey's rule is established in [79]. Furthermore in [79], Jeffrey proofs that multi-step chaining with respect to a collection of single-event updates of independent events is commutative, i.e., Theorem 5.3. Correspondence between Richard C. Jeffrey and Rudolf Carnap concerning [79] can be found in [80]. In [81, 87] Jeffrey establishes the Bayesian framework of the logic of decision, including probability kinematics and Jeffrey conditionalization as integral parts. A derivation of Jeffrey conditioning is also provided in [82], together with a succinct explanation of its epistemological foundations in the subjectivism of Frank P. Ramsey. In [88–90] Jeffrey re-established and refines the epistemological foundation of the logic of decision and probability kinematics as so-called radical probabilism. For a discourse concerning probability kinematics with Isaac Levi see [102] and Jeffrey's response in [83]. For a discussion of objectivism by Richard C. Jeffrey see [91].

Contingency Tables for Given Observations and Marginals

An important dual problem to partial conditionalization is about determining the most likely probability distribution with known marginals for a complete set of observations. This problem is treated by W. Edwards Deming and Frederick F. Stephan in [30] and C.T. Ireland and Solomon Kullback in [74]. Given two partitions of events $B_1, \ldots, B_s$ and $C_1, \ldots, C_t$ and absolute numbers of observations n_{ij} for all possible $B_i C_j$ in a sample of size n and furthermore given marginals $p_{i\star}$ for each B_i in and $p_{\star j}$ for each C_j, it is the intention to find a probability distribution P that adheres to the specified marginals, i.e., such that $P(B_i) = p_{i\star}$ for all B_i and $P(C_j) = p_{\star j}$ for all C_j, and furthermore maximizes the probability of the specified joint observation, i.e., that maximizes the following multinomial distribution; compare with Def. 2.11:

$$\mathfrak{M}_{n, P(B_1C_1), \ldots, P(B_1C_t), \ldots, P(B_sC_1), \ldots, P(B_sC_t)}(n_{11}, \ldots, n_{1t}, \ldots, n_{s1}, \ldots, n_{st})$$

© The Author(s) 2017
D. Draheim, *Generalized Jeffrey Conditionalization*, SpringerBriefs in Computer Science, https://doi.org/10.1007/978-3-319-69868-7

Note that the colleciton of $s \times t$ events $B_s B_t$ form a partition. The observed values n_{ij} are said to be organized in a two-dimensional $s \times t$ contingency table. The restriction to two-dimensional contingency tables is without loss of generality, i.e., the results of [30] and [74] can be generalized to multi-dimensional tables. In comparisons with partial conditionalizations, we treat two events B and C as a 2×2 contingency table with partitions $B_1 = B$, $B_2 = \overline{B}$, $C_1 = C$ and $C_2 = \overline{C}$. Now, [30] approaches the optimization by least-square adjustment, i.e., by considering the probability function P that minimizes $\chi^2 = \sum_{i=1}^{s} \sum_{j=1}^{t} (n_{ij} - n \cdot \mathsf{P}(B_i C_j))^2 / n_{ij}$, whereas [74] approaches the optimization by considering the probability function P that minimizes the Kullback-Leibler number $I(\mathsf{P}, \mathsf{P}') = \sum_{i=1}^{s} \sum_{j=1}^{t} \mathsf{P}(B_i C_j) \cdot \ln\big(\mathsf{P}(B_i C_j)/\mathsf{P}'(B_i C_j)\big)$ with $\mathsf{P}'(B_i C_j) = n_{ij}/n$; compare also with [98, 99]. Both [30], see, in particular, [141], and [74] use iterative procedures that generates BAN (best approximatively normal) estimators for convergent computations of the considered minima; compare also with [117, 144].

Rationales of Probability Kinematics, Closeness

Assessments of Jeffrey's radical probabilism [88–90] are provided by Richard Skyrms in [137] and Richard Bradley in [16]. A characterization of the essence and at the same time generalization of the probability kinematics postulate to the categories of responsiveness and conservativeness is achieved by Richard Bradley in [18]; compare also with Dietrich, List and Bradley in [36]. A mathematical analysis of Jeffrey conditionalization is provided by Persi Diaconis and Sandy Zabell in [33]. Also, [33] considers candidates for generalized Jeffrey conditionalization. The first is based on the assignment of the product probability $\mathsf{P}(A|\mathbf{B}) \cdot \mathsf{P}(A|\mathbf{C})$ for the conditionalization $\mathsf{P}(A|\mathbf{B}, \mathbf{C})$ with update vectors $\mathbf{B}$ and $\mathbf{C}$ that both individually form partitions. The second selects as conditionalization $\mathsf{P}_\mathbf{B}$ the probability distribution Q that adheres to the frequency specifications from the update vector $\mathbf{B}$ and minimizes the Kullback-Leibler number $I(\mathsf{Q}, \mathsf{P})$; compare with the definition above on p. 82, with the result that the *a priori* probability distribution P serves as observation in a sample and this way is moved into the *a posteriori* position, see also [27, 28] and again [30, 74, 141]. Further treatments and assessments of conditionalization via notions of closeness can be found in [112, 113] by Sherry May and William Harper, in [148] by Damjan Škulj and in [160] by Chunlai Zhou et al., where the latter is based on further developing the rules introduced by Philippe Smets in [138].

Non-Commutativity of Jeffrey Conditionalization

An experience-parameterized reformulation of Jeffrey conditionalization that aims at a consistent supplement with Bayesian input laws is provided by Harty Field in [57]. An assessment of Field's parameterized extension of Jeffrey conditionalization from [57] is provided by Daniel Garber in [66]. The correspondence between general Bayesian conditionalization, Jeffrey conditionalization and Fields's extension is analyzed by Zoltan Domotor in [37]. A systematic analysis of non-commutativity of Jeffrey conditionalization is provided by Bas van Fraassen in [61]. In [26] David Christensen transcends the discourse about the non-commutativity of Jeffrey conditionalization to the issue of confirmational holism. An assessment of the non-commutativity of Jeffrey conditionalization that is streamlined by the aim to meet rational psychology is provided Frank Döring in [39]. A review of the non-commutativity of Jeffrey conditionalization from the perspective of sensory experiences is provided by Mark Lange in [101]. For a systematic argumentation in favor for the non-commutativity of Jeffrey conditionalization see [149] by Carl G. Wagner. In [154] Jonathan Weisberg achieves to streamline the arguments of David Christensen concerning confirmational holism from [26] into a systematic rationale, see also [156]. A reassessment of Weisberg's analysis from [154] with respect to options in weakening rigidity assumptions is provided by Lydia McGrew in [114]. A reassessment of Weisberg's analysis from [154] with respect to theory-dependent evidence resulting in a concrete proposal on how to incorporate theory-dependent evidence in updating is provided by J. Dmitri Gallow in [65].

Dempster-Shafer Theory, Maximal Entropy Principle

A characterization of Jeffrey conditionalization in Dempster-Shafer theory [31, 32, 132] is provided by Glenn Shafer in [133, 134]. A comparative analysis of Bayes' rule, Jeffrey conditionalization and Dempster's rule is provided by Persi Diaconis and Sandy Zabell in [34, 35]. A synthesis of Dempster-style lower bounds with Jeffrey conditionalization is provided by Carl G. Wagner in [151, 152]. Generalizations of Jeffrey's rule that allow for the asymmetric treatment of evidence are provided by Hidetomo Ichihashi and Hideo Tanaka in [73]. The design of the rules in [73] are oriented towards the findings of Didier Dubois and Henri Prade [55] concerning Dempster's rule. A characterization of Jeffrey conditionalization in terms of the maximum entropy principle [75, 100] is provided by Brian Skyrms in [136]. An analysis of the interpretation of Jeffrey conditionalization in terms of the entropy principle together with a reassessment of Dempster-Shafer conditionals in term of this interpretation is provided by Paolo Garbolino in [67]. An analysis that also incorporates a systematic reassessment of Wagner's generalization of Jeffrey conditionalization is provided by Stefan Lukits in [107].

Epistemology, Inductive Reasoning, Dutch Book Arguments, Counterfactual Probabilities, Preference Orderings, Evidence Pooling, CAR, Representation Theorems, Possibility Theory, Rational Disagreement

In [146, 147] Paul Teller shows how the usage of conditionalization as a model of changing belief can be justified in terms of inductive reasoning. In [3] Brad Armendt deals with the establishment of a Dutch book argument for Jeffrey conditionalization. A critical examination of Dutch book arguments for ordinary and Jeffrey conditionalization is provided by Colin Howson in [71]. In [62] Bas Van Fraasen develops a generalized framework of epistemic judgments in which Jeffrey conditionalization fits as an instance. In [12] Craig Boutilier conducts an analysis of

counterfactual probability functions [139] that exploits Jeffrey-rule-style conditionalization. In [13] Richard Bradley investigates the correspondence between rational
preference orderings in Jeffrey's logic of decision and Ernest W. Adams' logic of
conditionals [1]. An analysis of Jeffrey conditionalization in terms of a generalized
CAR (coarsening at random) condition is provided by Peter D. Grünwald and Joseph
Y. Halpern in [126]. For an alternative measure-theoretic reconstruction of Ramsey's
representation theorem [128] and its application to Richard C Jeffrey's assessment
of Ramsey's representation theorem, see [15] by Richard Bradley. For the exploita-
tion of Jeffrey conditionalization for implicit feedback in information retrieval systems see [157] by Ryen W. White et al. In [150] Carl Wagner establishes a notion
of pooling individual evidence in chained conditionalizations via the notion of external Bayesianism [109] so that it falls together with Jeffrey conditionalization. A
semantics of Jeffrey conditionalization in Lofti Zadeh's possibility theory [56, 159]
is provided by Salem Benferhat et al. in [6–8]. An analysis of sustained rational dis-
agreement in terms of Jeffrey conditionalization is provided by Simon M. Hutegger
in [72]. A systematic characterization of the pragmatics of Jeffrey conditionalization
in decision scenarios is provided by Christopher J. G. Meacham in [115]. In [142]
Rush Stewart and Ignacio Ojea Quintana generalize Wagner's pooling from [150]
further so that they can exploit it in categorizing Bayesian updates with respect to
the notion of convex imprecise probability pooling that they introduce in [143].

Jeffrey Desirabilities

For the sake of completeness, we give two equivalent original definitions of desirability by Richard C. Jeffrey from [92] and [87]:

> "To the option of making the proposition A true corresponds the conditional proba-
> bility distribution $pr(\cdots |A)$, where the unconditional distribution $pr(\cdots)$ represents
> your prior probability judgment – prior, that is, to deciding which option-proposition
> to make true. And your expectation of utility associated with the A-option will be   (A.1)
> your conditional expectation $ex(u|A)$ of the random variable u (for "utility"). This
> conditional expectation is also known as your *desirability* for truth of A, and denoted
> 'desA': $desA = ex(u|A)$" [92], p. 102

> "The rule is this: *the desirability of a proposition is a weighted average of the de-*
> *sirabilities of the cases in which it is true, where the weights are proportional to the*
> *probabilities of the cases.* More explicitly, the procedure for computing the desir-
> ability of a proposition can be given in three steps: (1) multiply the desirability of   (A.2)
> each case in which the proposition is true by the probability of that case; (2) add all
> such products; and (3) divide by the probability of the proposition. (In other words,
> divide by the sum of the probabilities of the cases in which the proposition is true.)"
> [87], p. 78

In [14] Richard Bradley analyses desirabilities *a posteriori* from the perspective of
rational conditional preference, following the lines of Bolker's representation theo-
rem [9]. Based on the analysis, a new concept of conditional desirability $des(A|B)$
is introduced and defined as $des(A|B) = des(AB) - des(A) + des(\top)$; compare also
with [140]. The new conditional desirability $des(A|B)$ aims at filling a gap in the
deliberation framework and it is different from the corresponding Jeffrey desir-

ability *a posteriori*, which is $E_{P_{B\equiv100\%}}(v|A)$, which equals $E_{P_B}(v|A)$, which equals $E_P(v|AB)$, which equals $E(v|AB)$. Note that it is $E_{P_{B\equiv100\%}}(v|A) = E_{P_B}(v|A)$ that is symbolized by Jeffrey in implicit notation as $DES(A)$ in his discussion of *a posteriori* desirabilities on p. 90 in [87], where it is important that $B \equiv 100\%$ is maintained in the context and $DES(A)$ can be read only in the context of this special case; compare also with the discussion in Sect. 4.6 and 5.3, in particular, with rule (C) in Table 4.2 and Eqns. (5.17), (5.18), (5.19) and (5.20). In [17] Richard Bradley develops the results from [14] further into a coherent deliberation framework. In [20] Richard Bradley and H. Orri Stefánsson develop a model of counterfactual desirabilities resulting into a multidimensional possible-world semantics for conditionals. In [19] Richard Bradley and Christian List provide an assessment of David Lewis' desire-of-belief thesis [105, 106]. In [122] Ittay Nissan-Rozen investigates the preservation of David Lewis' principal principle [104] under Jeffrey conditionalization.

Conditional Probabilities and Conditional Events

A comprehensive treatment of conditional probabilities is provided by Malempati M. Rao in [130]. An axiomatic treatment of conditional probabilities is provided by Bernard O. Koopman in [97]. An axiomatic treatment of conditional probabilities, the construction and investigation of the conditional probability space and the establishment of a conditional law of large numbers on the basis of the constructed conditional probability space is provided by Alfréd Rényi in [131]. An analytical treatment of conditional probabilities is provided by David Lewis in [103]. A treatment of conditional events and conditional event algebras is provided by Irwin R. Goodman, Hung T. Nguyen and Elbert A. Walker in [68], see also [119]. An algebraic treatment of conditional events that mitigates between the measure-theoretic and the analytical viewpoint is provided by Tommaso Flaminio et al. in [60]. An experimental investigation of the cognitive aspects of conditional events is provided by Andrew J.B. Fugard et al. in [63].

Interpretations of Probability Theory

For treatments on the interpretations of probability theory see Julian Jaynes in [78], Appendix A and Appendix B, John Maynard Keynes in [93], Chapter VIII, in general, and Andrey Kolmogorov in [96], Jerzy Neyman in [118] and Alan Hájek in [69], in particular.

Appendix B
Basic Formulary and Notation

Definition B.1 (σ-Algebra) Given a set Ω, a σ-*Algebra* Σ *over* Ω is a set of subsets of Ω, i.e., $\Sigma \subseteq \mathbb{P}(\Omega)$, such that the following conditions hold true:

1) $\Omega \in \Sigma$
2) If $A \in \Sigma$ then $\Omega \backslash A \in \Sigma$
3) For all countable subsets of Σ, i.e., $A_0, A_1, A_2 \ldots \in \Sigma$ it holds true that $\bigcup\limits_{i \in \mathbb{N}} A_i \in \Sigma$

Definition B.2 (Probability Space) A *probability space* $(\Omega, \Sigma, \mathsf{P})$ consists of a set of outcomes Ω, a σ-algebra of (random) events Σ over the set of outcomes Ω and a probability function $\mathsf{P} : \Sigma \to \mathbb{R}$, also called probability measure, such that the following axioms hold true:

1) $\forall A \in \Sigma \,.\, 0 \leqslant \mathsf{P}(A) \leqslant 1$ (i.e., $\mathsf{P} : \Sigma \to [0,1]$)
2) $\mathsf{P}(\Omega) = 1$
3) (Countable Additivity): For all countable sets of pairwise disjoint events, i.e., $A_0, A_1, A_2 \ldots \in \Sigma$ with $A_i \cap A_j = \emptyset$ for all $i \neq j$, it holds true that

$$\mathsf{P}\Big(\bigcup_{i=0}^{\infty} A_i\Big) = \sum_{i=0}^{\infty} \mathsf{P}(A_i)$$

Definition B.3 (Measurable Space, Measurable Function) Given two *measurable spaces* (X, Σ) and (Y, Σ'), i.e., sets X and Y equipped with a σ-algebra Σ over X and a σ-algebra Σ' over Y. A function $f : X \to Y$ is called a *measurable function*, also written as $f : (X, \Sigma) \to (Y, \Sigma')$, if for all sets $U \in \Sigma'$ we have that the inverse image $f^{-1}(U)$ is an element of Σ.

Definition B.4 (Random Variable) A *random variable* X based on a probability space $(\Omega, \Sigma, \mathsf{P})$ is a measurable function $X : (\Omega, \Sigma) \to (I, \Sigma')$ with so-called *indicator set* I. The notation $(X = i)$ is used to denote the inverse image $X^{-1}(i)$ of an element $i \in I$ under f. It is usual to omit the σ-algebras in the definition of concrete random variables $X : (\Omega, \Sigma) \to (I, \Sigma')$ and specify them in terms of functions $X : \Omega \to I$ only. A random variable $X : \Omega \to I$ is called a *discrete random variable* if $X^{\dagger}(\Omega)$ is at most countable infinite.

© The Author(s) 2017

D. Draheim, *Generalized Jeffrey Conditionalization*, SpringerBriefs
in Computer Science, https://doi.org/10.1007/978-3-319-69868-7

Definition B.5 (Expected Value)   Given a real-valued discrete random variable $X : \Omega \to I$ with indicator set $I = \{i_0, i_1, i_2, \ldots\} \subseteq \mathbb{R}$ based on $(\Omega, \Sigma, \mathsf{P})$, the *expected value* $\mathsf{E}(X)$, or *expectation* of X (where E can also be denoted as E_P in so-called explicit notation) is defined as follows:

$$\mathsf{E}(X) = \sum_{n=0}^{\infty} i_n \cdot \mathsf{P}(X = i_n) \tag{B.1}$$

Definition B.6 (Conditional Expected Value) Given a real-valued discrete random variable $X : \Omega \to I$ with indicator set $I = \{i_0, i_1, i_2, \ldots\} \subseteq \mathbb{R}$ based on a probability space $(\Omega, \Sigma, \mathsf{P})$ and an event $A \in \Sigma$, the *expected value* $\mathsf{E}(X)$ *of X conditional on A* (where E can also be denoted as E_P in so-called explicit notation) is defined as follows:

$$\mathsf{E}(X|A) = \sum_{n=0}^{\infty} i_n \cdot \mathsf{P}(X = i_n \,|\, A) \tag{B.2}$$

Definition B.7 (Fractions) Each rational number $q \in \mathbb{Q}$ is represented by an *irreducible fraction n/d* with $q = n/d$ that consists of a numerator $n \in \mathbb{N}_0$ and a denominator $d \in \mathbb{N}$ such that n and d are minimal, i.e., cannot be divided further by any natural number $i \neq 1$. Given a list of natural numbers $(n_i)_{i \in \{1, \ldots, m\}}$ their *least common multiply* is defined as the least natural number $k \in \mathbb{N}$ such that $k/n_i \in \mathbb{N}$ and for all $1 \leqslant i \leqslant m$. Given a list of irreducible fractions $\mathbf{n} = (n_i/d_i)_{i \in \{1, \ldots, m\}}$ their *least common denominator*, denoted by $lcd(\mathbf{n})$ is defined as the least common multiply of the fraction denominators $(d_i)_{i \in \{1, \ldots, m\}}$. Given a list of rational numbers $\mathbf{q}$ with the list of irreducible numbers $\mathbf{n} = \mathbf{q}$, the *least common denominator* of $\mathbf{q}$, denoted by $lcd(\mathbf{q})$, is defined as the least common denominator $lcd(\mathbf{n})$.

Definition B.8 (Set and Function Notation)   Given a set $S \subseteq U$ such that U is assumed as known from the context, we use $\overline{S}$ to denote the *complement of S (in U)*, i.e., $\overline{S} = U \backslash S$. Given sets S and U such that $S \subseteq U$ is assumed as known from the context, we use $u \notin U$ to denote $u \in U \backslash S$. We use both $|S|$ and $\#S$ to denote the *set size* of S. Given a function $f : A \longrightarrow B$, we use $f^\dagger : \mathbb{P}(A) \longrightarrow \mathbb{P}(B)$ to denote its *lifted function*, i.e., $f^\dagger(A') = \{f(a) \,|\, a \in A'\}$ for each $A' \subseteq A$. Given a function $f : A \longrightarrow B$, we use $f^{-1} : B \longrightarrow \mathbb{P}(A)$ to denote its *inverse function*, i.e., $f^{-1}(b) = \{a \,|\, a \in A, \, f(a) = b\}$ for each $b \in B$.

Definition B.9 (Partition) Given a set M, a countable collection $B_1, B_2 \ldots$ of subsets of M is called a *partition (of M)* if $B_i \cap B_j = \emptyset$ for all $i \neq j$ and $\cup \{B_i \,|\, i \geqslant 1\} = M$.

Appendix C
Technical Lemmas and Proofs

Lemma C.1 (I.I.D. Multivariate Random Variable Independencies) *Given a sequence of i.i.d. multivariate random variables* $(\langle X_1,\ldots,X_n\rangle_i)_{i\in\mathbb{N}}$, *a finite set* $C\subset\mathbb{N}$ *of column indices and a set* $R_c\subseteq\{1,\ldots,n\}$ *of row indices for each* $c\in C$. *Then, for all families of index values* $((i_{\rho\kappa}:I_\rho)_{\rho\in R_\kappa})_{\kappa\in C}$ *we have that the following column-wise independence holds:*

$$P\left(\bigcap_{c\in C}\bigcap_{r\in R_c} X_{rc}=i_{rc}\right)=\prod_{c\in C}P\left(\bigcap_{r\in R_c} X_{rc}=i_{rc}\right) \tag{C.1}$$

Proof. We start with the left-hand side of Eqn. (C.1), i.e.,

$$P\left(\bigcap_{c\in C}\bigcap_{r\in R_c} X_{rc}=i_{rc}\right) \tag{C.2}$$

Henceforth, we denote the complement of R_c in $\{1,\ldots,n\}$ as $\bar{R}_c$ for each $c\in C$, i.e., $\bar{R}_c=\{1,\ldots,n\}\backslash R_c$. For each column $c\in C$ we augment the intersection of the conjuncts $X_{rc}=i_{rc}$ in Eqn. (C.2) by respective conjuncts for all rows not included in R_c. To meet the original marginal value of Eqn. (C.2), we must sum up over all possible combinations of data for the newly introduced rows resulting in Eqn. (C.3). Note that $((j_{\rho\kappa}:I_\rho)_{\rho\in\bar{R}_\kappa})_{\kappa\in C}$ in Eqn. (C.3) varies over all possible families of index values for all $\kappa\in C$ and all $\rho\in\bar{R}_\kappa$:

$$\sum_{((j_{\rho\kappa}:I_\rho)_{\rho\in\bar{R}_\kappa})_{\kappa\in C}} P\left(\bigcap_{c\in C}\left(\bigcap_{r\in R_c} X_{rc}=i_{rc},\bigcap_{r\in\bar{R}_c} X_{rc}=j_{rc}\right)\right) \tag{C.3}$$

Now, let us choose an arbitrary but fixed column $\xi\in C$ and single it out so that Eqn. (C.3) results into

$$\sum_{((j_{\rho\kappa}:I_\rho)_{\rho\in\bar{R}_\kappa})_{\kappa\in C}} P\left(\left(\bigcap_{r\in R_\xi} X_{r\xi}=i_{r\xi},\bigcap_{r\in\bar{R}_\xi} X_{r\xi}=j_{r\xi}\right),\bigcap_{c\in C\backslash\{\xi\}}\left(\bigcap_{r\in R_c} X_{rc}=i_{rc},\bigcap_{r\in\bar{R}_c} X_{rc}=j_{rc}\right)\right)$$
$$\tag{C.4}$$

© The Author(s) 2017

D. Draheim, *Generalized Jeffrey Conditionalization*, SpringerBriefs
in Computer Science, https://doi.org/10.1007/978-3-319-69868-7

Now, we can exploit that $(\langle X_1,\ldots,X_n\rangle_i)_{i\in\mathbb{N}}$ forms a sequence of i.i.d. multivariate random variables. Due to the respective independencies and exploiting Eqn. (2.12) we have that Eqn. (C.4) equals

$$
\sum_{((j_{\rho\kappa}:I_\rho)_{\rho\in\overline{R}_\kappa})_{\kappa\in C}} \Bigg(\ \mathsf{P}\Big(\underset{r\in R_\xi}{\cap}X_{r\xi}=i_{r\xi},\ \underset{r\in\overline{R}_\xi}{\cap}X_{r\xi}=j_{r\xi}\Big)
$$
$$
\times\mathsf{P}\Big(\underset{c\in C\backslash\{\xi\}}{\cap}\big(\underset{r\in R_c}{\cap}X_{rc}=i_{rc},\ \underset{r\in\overline{R}_c}{\cap}X_{rc}=j_{rc}\big)\Big)\Bigg)   \quad\text{(C.5)}
$$

As a next step we can expand the sum in Eqn. (C.5) so that it results into

$$
\sum_{((\iota_\rho:I_\rho)_{\rho\in R_\xi})}\Bigg(\ \mathsf{P}\Big(\underset{r\in R_\xi}{\cap}X_{r\xi}=i_{r\xi},\ \underset{r\in\overline{R}_\xi}{\cap}X_{r\xi}=\iota_r\Big)
$$
$$
\times\sum_{((j_{\rho\kappa}:I_\rho)_{\rho\in R_\kappa})_{\kappa\in C\backslash\{\xi\}}}\mathsf{P}\Big(\underset{c\in C\backslash\{\xi\}}{\cap}\big(\underset{r\in R_c}{\cap}X_{rc}=i_{rc},\ \underset{r\in\overline{R}_c}{\cap}X_{rc}=j_{rc}\big)\Big)\Bigg)
$$
$$
\text{(C.6)}
$$

Next, we can place the second factor in Eqn. (C.6) outside the brackets. Next, we can erase all particels X_{rc} for rows $r\in\overline{R}_c$ for some $c\in C$ and this way eliminate the sums so that Eqn. (C.6) results into

$$
\mathsf{P}\Big(\underset{r\in R_\xi}{\cap}X_{r\xi}=i_{r\xi}\Big)\times\mathsf{P}\Big(\underset{c\in C\backslash\{\xi\}}{\cap}\big(\underset{r\in R_c}{\cap}X_{rc}=i_{rc}\big)\Big)   \quad\text{(C.7)}
$$

Finally, we can apply the transformation from Eqn. (C.2) to Eqn. (C.7) as often as $|C|-1$ times starting from Eqn. (C.2) which yields Eqn. (C.1). $\square$

Corollary C.2 (I.I.D. Multivariate Random Variable Independencies)
Given a sequence of i.i.d. multivariate random variables $(\langle X_1,\ldots,X_n\rangle_i)_{i\in\mathbb{N}}$ such that $X_1,\ldots,X_n$ are mutually independent, we have that the following holds for each index set of tuples $I\subseteq\mathbb{N}\times\mathbb{N}$:

$$
\mathsf{P}(\bigcap_{\langle i,j\rangle\in I}X_{ij}))=\prod_{\langle i,j\rangle\in I}\mathsf{P}(X_{ij})   \quad\text{(C.8)}
$$

Proof. Due to the premise that $X_1,\ldots,X_n$ are themselves independent, Eqn. (C.8) follows immediately as a corollary from Lemma C.1. $\square$

Lemma C.3 (Identical Probabilities of Target Event Repetitions)   *Given an F.P. conditionalization $\mathsf{P}^n(A\mid B_1\equiv b_1,\ldots,B_m\equiv b_m)$ we have that the probability of $A_{(\sigma)}$ conditional on the given frequency specification is equal for all repetitions $1\leqslant\sigma\leqslant n$, i.e., we have for some value v:*

$$
\mathsf{P}(A_{(\sigma)}\mid\overline{B_1}^n=b_1,\ldots,\overline{B_m}^n=b_m)=v   \quad\text{(C.9)}
$$

Proof. Henceforth, let us denote the index set of $B_1,\ldots,B_m$ as $I = \{1,\ldots,m\}$. We have that the left-hand side of Eqn. (C.9) equals

$$\underbrace{P(A_{(\sigma)}, \cap_{i\in I} B_i^n = b_i n)}_{\mu_\sigma} / P(\cap_{i\in I} B_i^n = b_i n) \tag{C.10}$$

It suffices to show that the numerator μ_σ of Eqn. (C.10) is equal for all σ. We have that μ_σ equals

$$P\Big(A_{(\sigma)}, \cap_{i\in I}\big(B_{i(1)} + \cdots + B_{i(m)} = b_i n\big)\Big) \tag{C.11}$$

First, we segment Eqn. (C.11) by decoupling each $B_{i(\sigma)}$ from the rest of the sum $B_{i(1)} + \cdots + B_{i(m)}$, simply according to the definition of sums of random variables. We do the decoupling by running a combined selection index $\langle l_1,\ldots,l_m\rangle$ over all feasible combinations, i.e., such that each l_i is either 0 or 1, yielding

$$\sum_{(l_i\in\{0,1\})_{i\in I}} P\Big(A_{(\sigma)}, \cap_{i\in I} B_{i(\sigma)} = l_i, \underset{\substack{1\leqslant\rho\leqslant n\\ \rho\neq\sigma}}{\cap_{i\in I} \textstyle\sum} B_{i(\rho)} = b_i n - l_i\Big) \tag{C.12}$$

Due to Lemma C.1 and the fact that $\rho \neq \sigma$ for the range index in Eqn. (C.12) we have that Eqn. (C.12) equals

$$\sum_{(l_i\in\{0,1\})_{i\in I}} \Big(P\big(A_{(\sigma)}, \cap_{i\in I} B_{i(\sigma)} = l_i\big) \cdot P\big(\underset{\substack{1\leqslant\rho\leqslant n\\ \rho\neq\sigma}}{\cap_{i\in I} \textstyle\sum} B_{i(\rho)} = b_i n - l_i\big)\Big) \tag{C.13}$$

Due to the fact that $P^n(A \mid B_1 \equiv b_1,\ldots,B_m \equiv b_m)$ is an F.P. conditionalization, we have that $(\langle A,B_1,\ldots,B_m\rangle_{(j)})_{j\in\mathbb{N}}$ is an i.i.d. sequence of multivariate characteristic random variables; compare with Def. 2.12. Therefore, we can apply Eqn. (2.17) to the second probability in Eqn. (C.13). Therefore, we have that Eqn. (C.13) equals

$$\sum_{(l_i\in\{0,1\})_{i\in I}} \underbrace{P\Big(A_{(\sigma)}, \cap_{i\in I} B_{i(\sigma)} = l_i\Big)}_{\phi_1} \cdot \underbrace{P\Big(\cap_{i\in I} B_i^{n-1} = b_i n - l_i\Big)}_{\phi_2} \tag{C.14}$$

Next, we consider ϕ_1 and ϕ_2 in Eqn. (C.14) for an arbitrary but fixed selection index $\langle l_1,\ldots,l_m\rangle$. Due to the fact that $(\langle A,B_1,\ldots,B_m\rangle_{(j)})_{j\in\mathbb{N}}$ is identically distributed, we have that ϕ_1 is the same for all σ. Furthermore, ϕ_2 is free from σ. Altogether, we have that the whole sum in Eqn. (C.14) is the same for all σ. $\qquad\square$

Lemma C.4 (Preservation of Independence under Aggregates) *Given m collections of real-valued random variables $X_{11},\ldots,X_{1n_1}$ through $X_{m1},\ldots,X_{mn_m}$ such that $X_{11},\ldots,X_{1n_1},\ldots,X_{m1},\ldots,X_{mn_m}$ are mutually independent, we have that the following holds true for all real numbers $k_1,\ldots,k_m$:*

$$P(X_1^{n_1} = k_1,\ldots,X_m^{n_m} = k_m) = P(X_1^{n_1} = k_1) \times \cdots \times P(X_m^{n_m} = k_m) \tag{C.15}$$

Proof. In order to proof Eqn. (C.15), it suffices to show the following:

$$P(X_1^{n_1}=k_1,\ldots,X_m^{n_m}=k_m) = P(X_1^{n_1}=k_1) \times P(X_2^{n_2}=k_2,\ldots,X_m^{n_m}=k_m) \quad \text{(C.16)}$$

We have that $P(X_1^{n_1}=k_1,\ldots,X_m^{n_m}=k_m)$ equals

$$\sum_{\substack{r_{11},\ldots,r_{1n_1},\ldots,r_{m1},\ldots,r_{mn_m} \in \mathbb{R} \\ \forall 1 \leqslant i \leqslant m.r_{i1}+\cdots+r_{in_i}=k_i}} P(X_{11}=r_{11},\ldots,X_{1n_1}=r_{1n_1},\ldots,X_{m1}=r_{m1},\ldots,X_{mn_m}=r_{mn_m})$$

$$\text{(C.17)}$$

Due to the premise that $X_{11},\ldots,X_{1n_1},\ldots,X_{m1},\ldots,X_{mn_m}$ are mutually independent we have that Eqn. (C.17) equals

$$\sum_{\substack{r_{11},\ldots,r_{1n_1},\ldots,r_{m1},\ldots,r_{mn_m} \in \mathbb{R} \\ \forall 1 \leqslant i \leqslant m.r_{i1}+\cdots+r_{in_i}=k_i}} P(X_{11}=r_{11},\ldots,X_{1n_1}=r_{1n_1}) \cdot P(X_{21}=r_{21},\ldots) \quad \text{(C.18)}$$

We have immediately that Eqn. (C.18) equals

$$\sum_{\substack{r_{11},\ldots,r_{1n_1} \in \mathbb{R} \\ r_{11}+\cdots+r_{1n_1}=k_1}} \left(P(X_{11}=r_{11},\ldots,X_{1n_1}=r_{1n_1}) \times \sum_{\substack{r_{22},\ldots,r_{2n_2},\ldots,r_{m1},\ldots,r_{mn_m} \in \mathbb{R} \\ \forall 2 \leqslant i \leqslant m.r_{i1}+\cdots+r_{in_i}=k_i}} P(X_{21}=r_{21},\ldots) \right) \quad \text{(C.19)}$$

We have immediately that Eqn. (C.19) equals

$$\sum_{\substack{r_{11},\ldots,r_{1n_1} \in \mathbb{R} \\ r_{11}+\cdots+r_{1n_1}=k_1}} P(X_{11}=r_{11},\ldots,X_{1n_1}=r_{1n_1}) \times \sum_{\substack{r_{22},\ldots,r_{2n_2},\ldots,r_{m1},\ldots,r_{mn_m} \in \mathbb{R} \\ \forall 2 \leqslant i \leqslant m.r_{i1}+\cdots+r_{in_i}=k_i}} P(X_{21}=r_{21},\ldots) \quad \text{(C.20)}$$

We have immediately that Eqn. (C.20) equals

$$P(X_1^{n_1}=k_1) \times P(X_2^{n_2}=k_2,\ldots,X_m^{n_m}=k_m) \quad \text{(C.21)}$$

With Eqn. (C.21) we have finished the proof of Eqn. (C.16). Having Eqn. (C.16), we can apply it $m-1$ times to $P(X_1^{n_1}=k_1,\ldots,X_m^{n_m}=k_m)$ yielding Eqn. (C.15). □

Lemma C.5 (Combining with Almost Impossible and Almost Sure Events)
Given a probability space (Ω,Σ,P) and events $A,B \in \Sigma$ we have the following:

$$P(B)=0 \implies P(AB)=0 \quad \text{(C.22)}$$
$$P(B)=1 \implies P(AB)=P(A) \quad \text{(C.23)}$$

Proof. With respect to Eqn. (C.22) we have that $AB \subseteq B$. Due to the monotonicity of P, i.e., $A \subseteq B \implies P(A) \leqslant P(B)$, we therefore immediately have Eqn. (C.22). With respect to Eqn. (C.23) we have that $P(A)$ equals $P(AB)+P(A\bar{B})$. With $P(B)=1$

we have that $P(\overline{B}) = 0$. Due to Eqn. (C.22) we therefore have that $P(A\overline{B} = 0)$ and therefore $P(A) = P(AB) + 0$, i.e., Eqn. (C.23).   □

Lemma C.6 (Law of Total Probabilities) *Given a probability space* (Ω, Σ, P), *an event* $A \subseteq \Omega$ *and a countable set of events* $B_1, B_2 \ldots$ *that form a partition of* Ω, *we have that*

$$P(A) = \sum_{i \geqslant 1} P(AB_i) \tag{C.24}$$

$$P(A) = \sum_{\substack{i \geqslant 1 \\ P(B_i) \neq 0}} P(B_i) \cdot P(A|B_i) \tag{C.25}$$

Proof. Basic law. Eqn. (C.24) holds due to the countable additivity of P, see Def. B.2, together with the premise that $B_1, B_2, \ldots$ form a partition. Eqn. (C.25) follows immediately from Eqn. (C.24), Lemma C.5 and the fact that $P(A|B_i)$ equals $P(AB_i)/P(B_i)$.   □

References

1. Adams EW (1975) The Logic of Conditionals – An Application of Probability to Deductive Logic. D. Reidel, Dordrecht, appeared as (Jaakko Hintikka, ed.): vol. 86 of Synthese Library
2. Aitken A, Gonin H (1935) On fourfold sampling with and without replacement. In: Proceedings of the Royal Society of Edinburgh, vol 55, pp 114–125
3. Armendt B (1980) Is there a dutch book argument for probability kinematics? Philosophy of Science 47:583–588
4. Atkinson C, Draheim D (2013) Cloud aided-software engineering – evolving viable software systems through a web of views. In: Mahmood Z, Saeed S (eds) Software Engineering Frameworks for the Cloud Computing Paradigm, Springer, Berlin Heidelberg, pp 255–281
5. Atkinson C, Draheim D, Geist V (2010) Typed business process specification. In: Proceedings of EDOC'2010 - the 14th IEEE International Enterprise Computing Conference, IEEE Press, pp 69–78
6. Benferhat S, Dubois D, Prade H, Williams M (2010) A framework for iterated belief revision using possibilistic counterparts to Jeffrey's rule. Fundamenta Informaticae 99(2):147–168
7. Benferhat S, Tabia K, Sedki K (2010) On analysis of unicity of Jeffrey's rule of conditioning in a possibilistic framework. In: ISAIM'10 – Proceedings of 11th International Symposium on Artificial Intelligence and Mathematics, Florida
8. Benferhat S, Tabia K, Sedki K (2011) Jeffrey's rule of conditioning in a possibilistic framework – an analysis of the existence and uniqueness of the solution. Annals of Mathematics and Artificial Intelligence 61:185–202
9. Bolker E (1966) Functions resembling quotients of measures. Transactions of the American Mathematical Society 124:292–312
10. Boole G (1854) The Laws of Thought – An Investigation of the Laws of Thought on Which Are Founded the Mathematical Theories of Logic and Probabilities. Walton and Maberly, Macmillan, London Cambridge
11. Bordbar B, Draheim D, Horn M, Schulz I, Weber G (2005) Integrated model-based software development, data access and data migration. In: Model Driven Engineering Languages and Systems, Springer, Berlin Heidelberg, Lecture Notes in Computer Science, vol 3713, pp 382–396
12. Boutilier C (1995) On the revision of probabilistic belief states. Notre Dame Journal of Formal Logic 36(1):158–183
13. Bradley R (1998) A representation theorem for a decision theory with conditionals. Synthese 116(2):187–229
14. Bradley R (1999) Conditional desirability. Theory and Decision 47(1):23–55
15. Bradley R (2004) Ramsey's representation theorem. Dialectica 58(4):483–498
16. Bradley R (2005) Radical probabilism and bayesian conditioning. Philosophy of Science 72(2):342–364

© The Author(s) 2017
D. Draheim, *Generalized Jeffrey Conditionalization*, SpringerBriefs
in Computer Science, https://doi.org/10.1007/978-3-319-69868-7

17. Bradley R (2007) The kinematics of belief and desire. Synthese – Bayesian Epistemology 156(3):513–535
18. Bradley R (forthcoming) Decision Theory with a Human Face. Draft (April 2016, pp. 318): http://personal.lse.ac.uk/bradleyr/pdf/Decision Theory with a Human Face (indexed 3).pdf
19. Bradley R, List C (2009) Desire-as-belief revisited. Analysis 69:31–7
20. Bradley R, Stefánsson HO (2017) Counterfactual desirability. The British Journal for the Philosophy of Science 68(2):485–533
21. Brentano F (1874) Psychologie vom empirischen Standpunkt. Duncker & Humblot, Leipzig
22. Brentano F (1995) Psychology from an Empirical Standpoint. Routledge, London New York
23. Carnap R (1945) The two concepts of probability – the problem of probability. Philosophy and Phenomenological Research 5(4):513–532
24. Carnap R (1962) Logical Foundations of Probability, 2nd edition. The University of Chicago Press
25. Chiarini A (2012) From Total Quality Control to Lean Six Sigma. Springer, Berlin Heidelberg, springerBriegs in Business
26. Christensen D (1992) Confirmational holism and bayesian epistemology. Philosophy of Science 59:540–557
27. Csiszár I (1967) Information type measures of difference of probability distributions and indirect observations. Studia Scientiarum Mathematicarum Hungarica 2:299–318
28. Csiszár I (1975) I-divergence geometry of probability distributions of minimization problems. Annals of Probability 3:146–158
29. Deming WE (2000) Out of the Crisis, 2nd edition. The MIT Press
30. Deming WE, Stephan FF (1940) On a least squares adjustment of a sampled frequency table when the expected marginal totals are known. The Annals of Mathematical Statistics 11(4):427–444
31. Dempster AP (1966) New methods for reasoning towards posterior distributions based on sample data. The Annals of Mathematical Statistics 37(2):355–374
32. Dempster AP (1968) A generalization of bayesian inference. Journal of the Royal Statistical Society, Series B 30(2):205–247
33. Diaconis P, Zabell S (1982) Updating subjective probability. Journal of the American Statistical Association 77(380):822–830
34. Diaconis P, Zabell S (1983) Some alternatives to bayes's rules. Tech. Rep. No. 205, Department of Statistics, Stanford University
35. Diaconis P, Zabell S (1986) Some alternatives to bayes's rules. In: Grofman B, Owen G (eds) Information Pooling and Group Decision Making, JAI Press, pp 25–38
36. Dietrich F, List C, Bradley R (forthcoming) Belief revision generalized – a joint characterization of Bayes's and Jeffrey's rules. Journal of Economic Theory
37. Domotor Z (1980) Probability kinematics and representation of belief change. Philosophy of Science 47(3):384–403
38. Donkin WF (1851) On certain questions relating to the theory of probabilities. Philosophical Magazine 1(5):353–368
39. Döring F (1999) Why bayesian psychology is incomplete. Philosophy of Science 66:379–389
40. Draheim D (2010) Business Process Technology – A Unified View on Business Processes, Workflows and Enterprise Applications. Springer, Berlin Heidelberg
41. Draheim D (2010) The service-oriented metaphor deciphered. Journal of Computing Science and Engineering 4(4):253–275
42. Draheim D (2012) Smart business process management. In: Fischer L (ed) 2011 BPM and Workflow Handbook, Digital Edition, Future Strategies, Workflow Management Coalition, pp 207–223
43. Draheim D (2016) Reflective constraint writing. Transactions on Large-Scale Data- and Knowledge-Centered Systems 24:1–60
44. Draheim D (2017) Semantics of the Probabilistic Typed Lambda Calculus – Markov Chain Semantics, Termination Behavior, and Denotational Semantics. Springer, Berlin Heidelberg

45. Draheim D, Weber G (2002) Strongly typed server pages. In: Proceedings of NGITS'2002 – the 5th International Workshop on Next Generation Information Technologies and Systems, Springer, Berlin Heidelberg, Lecture Notes in Computer Science, vol 2382, pp 29–44

46. Draheim D, Weber G (2003) Modeling submit/response style systems with form charts and dialogue constraints. In: Proceedings of HCISWWA 2003 – Workshop on Human Computer Interface for Semantic Web and Web Applications, Springer, Berlin Heidelberg, no. 2889 in Lecture Notes in Computer Science, pp 267–278

47. Draheim D, Weber G (2003) Storyboarding form-based interfaces. In: Rauterberf M, Menozzi M, Wesson J (eds) Proceedings of INTERACT 2003 – The 9th IFIP TC13 International Conference on Human-Computer Interaction, IOS Press, IFIP TC13, pp 343–350

48. Draheim D, Weber G (2005) Form-Oriented Analysis – A New Methodology to Model Form-Based Applications. Springer, Berlin Heidelberg

49. Draheim D, Weber G (eds) (2006) Post-Proceedings of TEAA'2006 – the VLDB Workshop on Trends in Enterprise Application Architecture, no. 3888 in Lecture Notes in Computer Science, Berlin Heidelberg

50. Draheim D, Weber G (eds) (2007) Post-Proceedings of TEAA'2007 – the 2nd International Conference on Trends in Enterprise Application Architecture, no. 4473 in Lecture Notes in Computer Science, Berlin Heidelberg

51. Draheim D, Horn M, Schulz I (2004) The schema evolution and data migration framework of the environmental mass database imis. In: Proceedings of SSDBM 2004 – the 16th International Conference on Scientific and Statistical Database Management, IEEE Press, pp 341–344

52. Draheim D, Lutteroth C, Weber G (2005) Generative programming for C#. ACM SIGPLAN Notices 40(8):29–33

53. Draheim D, Himsl M, Jabornig D, Leithner W, Regner P, Wiesinger T (2009) Intuitive visualization-oriented metamodeling. In: Proceedings of DEXA'2009 – the 20th International Conference on Database and Expert Systems Applications, Springer, no. 5690 in Lecture Notes in Computer Science, pp 727–734

54. Draheim D, Himsl M, Jabornig D, Küng J, Leithner W, Regner P, Wiesinger T (2010) Concept and pragmatics of an intuitive visualization-oriented metamodeling tool. Journal of Visual Languages and Computing 21(4):157–170

55. Dubois D, Prade H (1986) On the unicity of dempster rule of combination. International Journal of Intellignet Systems 1(2):133–142

56. Dubois D, Prade H (1988) Possibility Theory. Plenum Press, New Yor

57. Field H (1978) A note on Jeffrey conditionalization. Philosophy of Science 45:361–367

58. de Finetti B (1964) Foresight – its logical laws, its subjective sources. In: Kyburg HE, Smokler HE (eds) Studies in Subjective Probability, Wiley

59. de Finetti B (2017) Theory of Probability – A Critical Introductory Treatment. John Wiley & Sons, first issued in 1975 as a two-volume work

60. Flaminio T, Godoa L, Hosni H (2015) On the algebraic structure of conditional events. In: Proceedings of ECSQARU'15 – the 13th European Conference on Symbolic and Quantitative Approaches to Reasoning with Uncertainty, Springer, Berlin Heidelberg, no. 9161 in Lecture Notes in Computer Science, pp 106–116

61. van Fraassen BC (1989) Laws and Symmetry. Clarendon Press, Oxford

62. Fraassen BCV (1980) Rational belief and probability kinematics. Philosophy of Science 47(2):165–187

63. Fugard AJ, Pfeifer N, Mayerhofer B, Kleiter GD (2011) How people interpret conditionals – shifts towards the conditional event. Journal of Experimental Psychology – Learning, Memory, and Cognition 37(3):635–648

64. Galavotti MC (2011) The modern epistemic interpretations of probability – logicism and subjectivism. In: Gabbay D, Hartmann S, Woods J (eds) Handbook of the History of Logic, Elsevier, vol 10, pp 153–203

65. Gallow JD (2014) How to learn from theory-dependent evidence, or: Commutativity and holism – a solution for conditionalizers. The British Journal for the Philosophy of Science 65(3):493–519
66. Garber D (1980) Field and Jeffrey conditionalization. Philosophy of Science 47(1):142–145
67. Garbolino P (1987) A comparison of some rules for probabilistic reasoning. International Journal of Man-Machine Studies 27:709–716
68. Goodman IR, Nguyen HT, Walker EA (1991) Conditional Inference and Logic for Intelligent Systems – A Theory of Measure-free Conditioning. North-Holland
69. Hájek A (1997) "mises redux" – redux: Fifteen arguments against finite frequentism. In: Costanini D, Galavotti MC (eds) Probability, Dynamics and Causality – Essays in Honour of Richard C. Jeffrey, Kluwer Academic Publishers, pp 69–88, appeared also in Erkenntnis, vol. 45, 1997, pp. 327–335
70. Himsl M, Jabornig D, Leithner W, Draheim D, Regner P, Wiesinger T, Küng J (2007) A concept of an adaptive and iterative meta- and instance modeling process. In: In DEXA'2007 – the 18th International Conference on Database and Expert Systems Applications, Springer, Berlin Heidelberg, no. 4653 in Lecture Notes in Computer Science, pp 519–528
71. Howson C (1997) Bayesian rules of updating. In: Costanini D, Galavotti MC (eds) Probability, Dynamics and Causality – Essays in Honour of Richard C. Jeffrey, Kluwer Academic Publishers, pp 55–68, appeared also in Erkenntnis, vol. 45, 1997, pp. 195–208
72. Hutegger SM (2015) Merging of opinions and probability kinematics. The Review of Symbolic Logic 8(4):611–648
73. Ichihashi H, Tanaka H (1989) Jeffrey-like rules of conditioning for the dempster-shafer theory of evidence. International Journal of Approximate Reasoning 3:143–156
74. Ireland CT, Kullback S (1968) Contingency tables with given marginals. Biometrika 55(1):179–188
75. Jaynes E (1957) Information theory and statistical mechanics. Physical Review, Series II 106(4):620–630
76. Jaynes E (1989) Papers on Probability, Statistics and Statistical Physics. Kluwer Academic Publishers, Dodrecht Boston London, edited by E.D. Rosenkranz
77. Jaynes ET (1968) Prior probabilities. IEEE Transactions on Systems Science and Cybernetics 4(3):227–41
78. Jaynes ET (2003) Probability Theory. Cambridge University Press
79. Jeffrey RC (1957) Contributions to the theory of inductive probability. PhD thesis, Princeton University
80. Jeffrey RC (1957) Contributions to the theory of inductive probability, phd thesis, with carnap correspondence. Tech. Rep. ASP.2003.02, Box 12, Folder 3, University of Pittsburgh
81. Jeffrey RC (1965) The Logic of Decision, 1st edition. McGraw-Hill, New York
82. Jeffrey RC (1968) Probable knowledge. In: Lakatos I (ed) The Problem of Inductive Logic, North-Holland, Amsterdam New York Oxford Tokio, pp 166–180
83. Jeffrey RC (1970) Dracula meets wolfman – acceptance vs. partial belief. In: Swain M (ed) Induction, Acceptance and Rational Belief, Synthese Library, vol 26, D. Reidel, Dodrecht, pp 157–185
84. Jeffrey RC (1975) Carnap's empiricism. In: Maxwell G, Jr RMA (eds) Induction, Probability, and Confirmation, University of Minnesota Press, Minnesota Studies in the Philosophy of Science, vol 6, pp 37–49
85. Jeffrey RC (1976) Axiomatizing the logic of decision. In: Archives of Scientific Philosophy, Special Collections Department, University of Pittsburgh, vol Box 14, Folder 18
86. Jeffrey RC (1983) Bayesianism with a human face. In: Earman J (ed) Testing Scientific Theories, University of Minnesota Press, pp 133–156
87. Jeffrey RC (1983) The Logic of Decision, 2nd edition. University of Chicago Press
88. Jeffrey RC (1985) Probability and the art of judgment. In: Achinstein P, Hannaway O (eds) Observation, Experiment and Hypotheses in Modern Physical Science, MIT Press, Cambridge, pp 95–126
89. Jeffrey RC (1992) Probability and the Art of Judgment. Cambridge University Press

90. Jeffrey RC (1992) Radical probabilism – prospectus for a user's manual. Philosophical Issues 2 – Rationality in Epistemology:193–204

91. Jeffrey RC (1997) Unknown probabilities – in memory of annemarie anrod shimony. In: Costanini D, Galavotti MC (eds) Probability, Dynamics and Causality – Essays in Honour of Richard C. Jeffrey, Kluwer Academic Publishers, pp 187–195, appeared also in Erkenntnis, vol. 45, 1997, pp. 327–335

92. Jeffrey RC (2004) Subjective Probability – the Real Thing. Cambridge University Press

93. Keynes JM (1921) A Treatise of Probability. Macmillan & Co., London

94. Kolmogorov A (1933) Grundbegriffe der Wahrscheinlichkeitsrechnung. Springer, Berlin Heidelberg

95. Kolmogorov A (1956) Foundations of the Theory of Probability. Chelsea

96. Kolmogorov A (1982) On logical foundation of probability theory. In: Itô K, Prokhorov JV (eds) Probability Theory and Mathematical Statistics, Springer, Berlin Heidelberg, Lecture Notes in Mathematics 1021, pp 1–5

97. Koopman BO (1940) The bases of probability. Bulletin of the American Mathematical Society 46(10):763–774

98. Kullback S (1959) Information Theory and Statistics. Wiley, New York

99. Kullback S, Khairat M (1966) A note on minimum discrimination information. The Annals of Mathematical Statistics 37:279–280

100. Kullback S, Leibler RA (1951) On information and sufficiency. The Annals of Mathematical Statistics 22(1):79–86

101. Lange M (2000) Is Jeffrey conditionalization defective by virtue of being non-commutative – remarks on the sameness of sensory experiences. Synthese 123:393–403

102. Levi I (1967) Probability kinematics. The British Journal for the Philosophy of Science 18(3):197–209

103. Lewis D (1976) Probabilities of conditionals and conditional probabilities. The Philosophical Review 85(3):297–315

104. Lewis D (1980) A subjectivist's guide to objective chance. In: Jeffrey RC (ed) Studies in Inductive Logic and Probabilities, University of California Press, Berkeley, vol 2, pp 263–293

105. Lewis D (1988) Desire as belief. Mind 97:323–32

106. Lewis D (1996) Desire as belief (ii). Mind 105:303–13

107. Lukits S (2015) Maximum entropy and probability kinematics constrained by conditionals. Entropy 17:1690–1700

108. Lutteroth C, Draheim D, Weber G (2011) A type system for reflective program generators. Journal Science of Computer Programming 76(5):392–422

109. Madansky A (1964) Externally bayesian groups. Tech. Rep. RM-4141-PR, RAND Corporation

110. Marshall AW, Olkin I (1983) A family of bivariate distributions generated by the bivariate bernoulli distribution. Tech. Rep. No. 187, Department of Statistics, Stanford University, Stanford

111. Marshall AW, Olkin I (1985) A family of bivariate distributions generated by the bivariate bernoulli distribution. Journal of the American Statistical Association 80(390):332–338

112. May S (1976) Probability kinematics – a constrained optimization problem. Journal of Philosophical Logic 5:395–398

113. May S, Harper W (1976) Toward an optimization procedure for applying minimum change principles in probability kinematic. In: Harper W, CAHooker (eds) Foundations of Probability Theory, Statistical Inference, and Statistical Theories of Science, vol. 1, The University of Western Ontario Series in Philosophy of Science, vol 6a, D. Reidel, Dordrecht, pp 137–166

114. McGrew L (2014) Jeffrey conditioning, rigidity, and the defeasible red jelly bean. Philosophical Studies 168(2):569–582

115. Meacham CJG (2015) Understanding conditionalization. Canadian Journal of Philosophy 45(5–6):767–797

116. Neyman J (1937) Outline of a theory of statistical estimation based on the classical theory of probability. Philosophical Transactions of the Royal Society of London 236(767):333–380
117. Neyman J (1946) Contribution to the theory of the x^2 test. In: Neyman J (ed) Proceedings of the Berkeley Symposium on Mathematical Statistics and Probability, University of California Press, Berkeley Los Angeles, pp 239–273
118. Neyman J (1977) Frequentist probability and frequentist statistics. Synthese 36:97–131
119. Nguyen HT, Walker EA (1994) A history and introduction to the algebra of conditional events and probability logic. IEEE Transactions on Systems, Man, and Cybernetics 24(12):1671–1675
120. Nilsson NJ (1986) Probabilistic logic. Artificial Intelligence 28:71–87
121. Nilsson NJ (1993) Probabilistic logic revisited. Artificial Intelligence 59:39–42
122. Nissan-Rozen I (2013) Jeffrey conditionalization, the principal principle, the desire as belief thesis, and adams's thesis. The British Journal for the Philosophy of Science 64:837–850
123. Pearl J (1988) Probabilistic Reasoning in Intelligent Systems – Networks of Plausible Inference, 2nd edition. Morgan Kaufmann, San Francisco
124. Pearl J (2009) Causal inference in statistics – an overview. Statistics Surveys 3:96–146
125. Pearl J (2009) Causality – Models, Reasoning, and Inference, 2nd edition. Cambridge University Press
126. Peter D Grünwald JYH (2002) Updating probabilities. In: Proceedings of UAI'02 – the 18th Conference in Uncertainty in Artificial Intelligence, AUAI Press, pp 187–196
127. Ramsey FP (1931) The Foundations of Mathematics and other Logical Essays. Kegan, Paul, Trench, Trubner & Co. Ltd., New York, edited by R.B. Braithwaite
128. Ramsey FP (1931) Truth and probability. In: Ramsey FP, Braithwaite R (eds) The Foundations of Mathematics and other Logical Essays, Kegan, Paul, Trench, Trubner & Co. Ltd., New York, pp 156–198
129. Ramsey FP (1990) Philosophical Papers. Cambridge University Press, edited by D.H. Mellor
130. Rao MM (2005) Conditional Measures and Applications, 2nd edition. Chapman and Hall, 1st edition published in 1993 by Marcel Dekker, New York
131. Rényi A (1995) On a new axiomatic theory of probability. Acta Mathematica Hungarica 6(3-4):285–335
132. Shafer G (1976) A Mathematical Theory of Evidence. Princeton University Press
133. Shafer G (1979) Jeffrey's rule of conditioning. Tech. Rep. 131, Department of Statistics, Stanford University
134. Shafer G (1981) Jeffrey's rule of conditioning. Philosophy of Science 48(3)
135. Skyms B (1986) Choice and Chance – An Introduction to Inductive Logic, 3rd edition. Wadsworth, Belmont
136. Skyrms B (1987) Updating, supposing and MAXENT. Theory and Decision 22:225–246
137. Skyrms B (1997) The structure of radical probabilism. In: Costanini D, Galavotti MC (eds) Probability, Dynamics and Causality – Essays in Honour of Richard C. Jeffrey, Kluwer Academic Publishers, pp 145–157, appeared also in Erkenntnis, vol. 45, 1997, pp. 285–297
138. Smets P (1993) Jeffrey's rule of conditioning generalized to belief functions. In: Heckerman D, Hamdani E (eds) Proceedings of UAI'93 – the 9th Conference on Uncertainty in Artificial Intelligence, AUAI Press, pp 500–505
139. Stalnaker RC (1981) Probability and conditionals. In: Harper WL, Stalnaker R, Pearce G (eds) Ifs – Conditionals, Belief, Decision, Chance and Time, D. Reidel Publishing Company, Dodrecht, Boston, The University of Western Ontario Series in Philosophy of Science, vol 15, pp 107–128
140. Stefánson HO (2014) Decision theory and counter-factual evaluation. PhD thesis, London School of Economics and Political Science
141. Stephan FF (1942) An iterative method of adjusting sample frequency tables when expected marginal totals are known. The Annals of Mathematical Statistics 13(2):166–178
142. Stewart R, Quintana IO (2017) Learning and pooling, pooling and learning. Erkenntnis 82:1–21

143. Stewart R, Quintana IO (2017) Probabilistic opinion pooling with imprecise probabilities. Journal of Philosophical Logic pp 1–29
144. Taylor WF (1953) Distance functions and regular best asymptotically normal estimates. The Annals of Mathematical Statistics 24(1):85–92
145. Teicher H (1954) On the multivariate poisson distribution. Skandinavisk Aktuarietidskrift 37:1–9
146. Teller P (1973) Conditionalization and observation. Synthese 26:218–258
147. Teller P (1976) Conditionalization and observation, and change of preference. In: Harper W, Hooker C (eds) Foundations of Probability Theory, Statistical Inference, and Statistical Inference of Science, vol. 2, D. Reidel, Dodrecht Boston, Philosophy of Science, Methodology, and Epistemology, vol 6, pp 205–253
148. Škulj D (2006) Jeffrey's conditioning rule in neighbourhood models. International Journal of Approximate Reasoning 42:192–211
149. Wagner C (2002) Probability kinematics and commutativity. Philosophy of Science 69:266–278
150. Wagner C (2009) Jeffrey conditioning and external bayesianity. Logic Journal of the IGLP 18(2):336–345
151. Wagner CG (1992) Generalized probability kinematics. Erkenntnis 36(2):245–257
152. Wagner CG (1992) Generalizing Jeffrey conditionalization. In: Dubois D, Wellman M, D'Ambrosio B, Smets P (eds) Proceedings of UAI'1992 – the 8th Conference on Uncertainty in Artificial Intelligence, Morgan Kaufmann, San Mateo, pp 331–335
153. Weirich P (2011) The bayesian decision-theoretic approach to statistics. In: Bandyopadhyay PS, Forster MR (eds) Philosophy of Statistics, North-Holland, Amsterdam Boston Heidelberg, (Dov M. Gabbay, Paul Thagard, John Woods, general editors) Handbook of Philosophy of Science, vol 7
154. Weisberg J (2009) Commutativity or holism – a dilemma for conditionalizers. The British Journal of the Philosophy of Sciences 60(4):793–812
155. Weisberg J (2011) Varieties of bayesianism. In: Gabbay D, Hartmann S, Woods J (eds) Handbook of the History of Logic, vol 10
156. Weisberg J (2014) Updating, undermining, and independence. The British Journal for the Philosophy of Science 66:121–159
157. White RW, Jose JM, van Rijsbergen CJ, Ruthven I (2004) A simulated study of implicit feedback models. In: McDonald S, Tait J (eds) Proceedings of ECIR'04 – the 26th European Conference on Information Retrieval, Springer, Berlin Heidelberg, Lecture Notes in Computer Science, vol 2997, pp 311–326
158. Wicksell SD (1916) Some theorems in the theory of probability – with special reference to their importance in the theory of homograde correlations. Svenska Aktuarieforeningens Tidskrift pp 165–213
159. Zadeh L (1978) Fuzzy sets as the basis for a theory of possibility. Fuzzy Sets and Systems 1:3–28
160. Zhou C, Wang M, Qin B (2014) Belief-kinematics Jeffrey's rules in the theory of evidence. In: Proceedings of UAI'14 – the 30th Conference on Uncertainty in Artificial Intelligence, AUAI Press, pp 917–926

Index

© The Author(s) 2017
D. Draheim, *Generalized Jeffrey Conditionalization*, SpringerBriefs
in Computer Science, https://doi.org/10.1007/978-3-319-69868-7

Printed in the United States
By Bookmasters